Mariya Nikolova
How Whiteness Claimed the Future

American Frictions

———

Editors
Carsten Junker
Julia Roth
Darieck Scott

Volume 7

Mariya Nikolova

How Whiteness Claimed the Future

The Always New vs The Always Now
in US-American Literature

DE GRUYTER

An earlier draft of this book was submitted as dissertation in partial fulfillment of the requirements for doctoral degree to Potsdam University in 2020 (supervisors: Prof. Nicole Waller, Prof. Sabine Broeck)

ISBN 978-3-11-163116-5
e-ISBN (PDF) 978-3-11-079999-6
e-ISBN (EPUB) 978-3-11-089133-1
ISSN 2698-5349

Library of Congress Control Number: 2022951411

Bibliographic information published by the Deutsche Nationalbibliothek
The Deutsche Nationalbibliothek lists this publication in the Deutsche Nationalbibliografie;
detailed bibliographic data are available on the Internet at http://dnb.dnb.de.

www.degruyter.com

Contents

Before chapter one

We try to catch up. The woods of Rila mountain are endless. Trapped in the trees, the wind as if sings and my mother lets her voice join, hiking ahead. She does not follow a path. My sister and I rush. We skip words from the song. We always hike like that: part myth, part masochism, singing. Only three times we stray away from each other. There is an old Bulgarian tradition of hugging trees, eyes closed. My mother has done it and my grandmother has and my great grandmother, too. My sister and I are young so do not question what the tradition is meant to protect. We know this: when you hike in the woods you will pass by a tree, feel it belongs to you, then you will go and hug it, blind. My mother says this is how you and the tree exchange energies. She says strength is secret – it happens in full darkness. No one can tell you which tree is *yours* and as kids we do not search for patterns or proof. We see our tree, run, wrap around it, and so we grow (there is a chance, of course, that we hug the same tree my grandmother has). After the embrace, we continue hiking and no one speaks of what they have told the tree or what the tree has told them in return. Rila seems heavy and hushed now but going forward is easier and my mother encourages us to sing as loud as we can. The daughters we are, we keep the tradition. It is only now I know what it means.

[jumpcut]

The book you hold in your hands started off as an analysis of the white American avant-garde. Interested in the ideological workings of fiction and following Toni Morrison and Frank Wilderson for the present interrogations of whiteness

> white people have a very, very serious problem and they should start thinking about what they can do about it (Morrison, "Interview")
> [among other things]
> a close reading on how Human capacities are manifest in specific genders and cultures (Wilderson, Afropessimism 333)

I wanted to study how major avant-garde tropes promote the potential of permanent renewal as American culture's (re)birth and transformation. The questions I had around the avant-garde and its cult of renewal had come pressing from my engagement with Black cultural critique. Renewal, in its broadest sense, is tied to the capacities to create, progress, transcend, and simply *be*, and Black critique (as I understand and conceptualize in the following) argues that, within dominant discourse, all these capacities have been stifled and denied to Black bodies ever since colonization. On the one hand, Black creative work and origin/ality have

https://doi.org/10.1515/9783110799996-001

been fetishized, appropriated, stolen, and dismissed in and by dominant culture (see Aldridge, Le Gall, Scafidi, Ziyad). On the other hand, Black being has been construed as negativity and barred on the level of ontology (see Fanon, Wilderson, Warren). It follows then that racialization operates on multiple levels in the conceptual frame of renewal.

My interest was to see how white texts engage and enact this racialization, whether they address it and in what way, but also how the texts "sound" if we read them with racialization as focus. I selected novels that not only openly thematize renewal but are celebrated for their transformational value, an impulse to expose past oppressions, and the subsequent rupture with aesthetic traditions leading to a new state (of being). Leaving aside the question of periodicity, I found that "avant-garde" is a useful common denominator in searching for these novels. At this point of the journey, I was following Susan Buck-Morss' understanding of avant-gardism, or the capacity to be avant-garde. For Buck-Morss, avant-gardism is not restricted to certain groups and timeframes. Rather, it denotes the "power of any cultural object to arrest the flow of history, and open up time for alternative visions." Buck-Morss writes,

> [w]hat counts is that the aesthetic experience teach us something new about our world, that it shock us out of moral complacency and political resignation, and that it take us to task for the overwhelming lack of social imagination that characterizes so much of cultural production in all its forms. (63)

In the meaning of pioneering, innovative, and socially relevant work, avant-gardism can be attributed to a wide range of experimental, rebellious, inspirational, and visionary fiction. Working with this broad definition, I depart from criticism which has discussed postmodern literature as a break with modernism and had thus left the avant-garde back in the late nineteenth and early twentieth century.[1] Instead, I assumed a *changing same* of modernism and postmodernism's programmatic quest for epistemic and aesthetic innovation. The latter move allows for a critique of white American culture's claim of self-renewal as systemic, an intrinsic property.

Selecting the novels for this study, I followed both critical reception and my intuition. I compiled a corpus of texts that were said to carry avant-gardist qualities and were explicit about their interest in and thematization of renewal. What I did not realize at the time, and what became clear to me in the process of writing this book, is that the avant-gardist qualities I and other critics were encountering were

---

1 see for instance Susan Sontag's *Styles of Radical Will*, Leslie Fiedler's "The New Mutants," and Ihab Hassan's *The Dismemberment of Orpheus: Toward a Postmodern Literature*

not so much linked to genre, formal techniques, or even taste. Rather, there was a condition of avant-gardism that came from the novels' whiteness. In other words, avant-gardism was read into certain works despite their lack of formal or conceptual innovation because the whiteness of these works, their status as written by white authors,[2] and their embrace of and embeddedness in white ideology, positioned them as socially significant, ground-breaking, and transformative.

One immediate clue for this occurred when I considered the texts' canonization. Although the selected novels were categorically different on a technical and subject level, they shared a similar reception. In fact, avant-gardism was often read into texts that were neither constructed nor counted as avant-garde. At this point, I became fascinated with tracing claims of avant-gardism, per Buck-Morss' definition, outside productions by the usual suspects – Gertrude Stein, Ezra Pound, T.S. Elliot, etc. *As Cathy Hong says, the avant-garde canon is exclusionary white so "[f]uck the avant-garde" ("Delusions").* I chose to consider Don DeLillo's *Zero K* (2016), Kathy Acker's *Kathy Goes to Haiti* (1978), and Marilynne Robinson's *Gilead* (2004) instead. Later, I understood that leaving the canon aside had permitted me to examine avant-gardism more critically, and that much of this critique – and its significance – would have remained curbed had my analysis stayed within the prescribed routes.

---

2 the designation "white" refers to a structural positionality throughout. In this, I follow Frank Wilderson's categorization (see Wilderson's *Red, White, and Black* and *Afropessimism*) where 'white' refers to anyone who occupies and may occupy the position of whiteness, structurally and performatively. This position enables and sustains both Euro-American white people and their 'junior partners' (to use Wilderson's term). Therefore, I do not make a distinction between diverse nuances/degrees within whiteness itself: e.g., my Bulgarian whiteness, Acker's Jewish whiteness, Bercovitch's Canadian whiteness, DeLillo's Italian whiteness, etc. Such distinction would go against Afropessimism's argument that whiteness, as a structural positionality and privilege, works for all non-Black people. Projects deriving from different assumptions have discussed the distinction (e.g., *Historicizing "Whiteness" in Eastern Europe and Russia* and the growing number of scholars who engage the topic, most recently Ivan Kalmar in *White But Not Quite: Race and Liberalism in Central Europe*). Personal experiences of discrimination and othering have convinced me that my whiteness diffracts and differs from central European whiteness, or American whiteness, or Australian whiteness (I will never forget one German professor's bewilderment when I positioned myself as a white reader for as he claimed, I should be aware that "in Germany, I am not *fully or really* white"). Albeit not the purview of my project, investigations of white-on-white violence and discriminatory practices against the *not fully or really* white people are certainly worthwhile. Violence and discrimination are harrowing experiences, whatever the enabling authority behind them and whoever their target. My interest lies in the workings of whiteness on the level of structure which is why I have forgone a more *intersectional* reading, and why I have, following Afropessimism, refused to rate or differentiate between degrees of whiteness.

Avant-gardism in whiteness. White ideology. White literature

Before I turn to the question of avant-gardism, I need to clarify a few important terms. I use whiteness and white ideology interchangeably. To me, the terms index the sum of beliefs, practices, norms, and expressions of white authority as constructed in dominant discourse from colonization until the present day. In this understanding, I follow Peter McLaren who explains that whiteness is a

> sociohistorical form of consciousness, given birth at the nexus of capitalism, colonial rule, and the emergent relationships among dominant and subordinate groups. Whiteness constitutes and demarcates ideas, feelings, knowledge, social practices, cultural formations, and systems of intelligibility that are identified or attributed to White people and that are invested in by White people as White. (149)

As we learn from Black and Indigenous critique, the investment of white authority has always been premised on eliminating the "subordinate groups" in McLaren's definition (see Fanon; Du Bois). Or, the logic and logistics of whiteness have always depended on the antagonism between white and non-white. "The concept of whiteness," explains Jess Row, "dates to the early seventeenth century, and originates alongside the enslavement of Africans for transport to the New World." At its core, Row writes further, "whiteness needs blackness to confirm that it is white," or whiteness depends on its antagonistic relationship to what it is not (46). The function, the operational coda, between the sides of this antagonism, is racialized violence coming from white people and directed at non-white people (see Fanon; Wilderson). This is how the calculation works; this is what births whiteness and protects it from disappearing.

It is important to mention here that despite the basics of this duality, whiteness is an open- ended term, for its own "internal mutation is limitless" (Wilderson, *Red, White, and Black* 297). In other words, certain procedures and applications of whiteness have changed over time – they have been criticised and either dismantled by Black and Indigenous[3] work or swept under the carpet by white

---

3 From here on I discuss US-American literature with regards to the antagonism whiteness – Blackness, i.e. I leave aside the tensions between white Americans and Native Americans, or the antagonism whiteness – Indigeneity. Because of Black people's displacement during the transatlantic slave trade, one often forgets that Black people are also Indigenous – to Africa. In other words, Indigeneity is contained in Black positionality even if this goes unnoticed. It should also be mentioned that throughout history, whiteness has positioned many different groups as Black, among them Indigenous populations. Or, Indigeneity and Blackness intersect, overlap, and reflect each other. As Wilderson argues, however, in the North American context this relation is neither simple, nor straightforward, and the positionalities are as connected as disparate at times (see Wilderson's

work. However, other procedures and applications come in their place because the underlaying structure and purpose of whiteness remains the same (see Wilderson, Sexton). In fact, whiteness's capacity to mutate limitlessly is what we can call an avant-gardism embedded in whiteness, or whiteness's programmatic quest towards renewal. The main purpose of this book is to analyse this capacity by looking closer at white literature.

My criteria for determining whether a text is derivate of whiteness is simple. Let's say, I examine a certain literary trope. If the trope effects the solidification of white authority, if it works for it, I consider it a part of whiteness's inventory. This method has its limitations – it is hard to grasp the extent of damage a trope inflicts on the imagination, and to list and pinpoint every single violation committed in its name. As Karen E. Fields and Barbara J. Fields suggest, such impossibility produces the "invisible ontology," or racecraft, through which whiteness and racialized violence are sustained over time. For them, racecraft refers to

> mental terrain and to pervasive belief. Like physical terrain, racecraft exists objectively; it has topographical features ... Unlike physical terrain, racecraft originates not in nature but in human action and imagination; it can exist in no other way. The action and imagining are collective yet individual, day-to-day yet historical, and consequential even though nested in mundane routine. The action and imagining emerge as part of moment-to-moment practicality, that is, thinking about and executing every purpose under the sun. (18–19)

If racecraft is the action, i.e. the work of racialized imagination and activity, whiteness is the authority requiring and enforcing this work. The longstanding traditions that have legitimized and thus rendered this authority normative disallow the halt or hold of it. Dominant discourse prohibits the very act of recognizing authority – even if the hurt whiteness inflicts is jarringly obvious, white authority frames its agents as anything but white. One reason for this is that "white people claim and achieve authority ... by not admitting, indeed not realising, that for much of the time [they act and think] only for whiteness" (Dyer xiv). In other words, whites experience themselves as "non-raced," as a quintessentially heterogenous group of individuals. As Richard Dyer explains, because whites are placed as the norm

---

Red, White, and Black). Interested in the ways white American literature uses Blackness to claim renewal, I limit my analysis and focus exclusively on the antagonism whiteness – Blackness. This choice produces an absence – and my study should be read with it in mind. On the topic of Black and Indigenous relationships, antagonisms, and solidarity see also King et al.'s *Otherwise Worlds. Against Settler Colonialism and Anti-Blackness*.

they seem not to be represented to themselves as white but as people who are variously gendered, classed, sexualised and abled. At the level of racial representation, in other words, whites are not of a certain race, they're just the human race. (3)

Responding to this misrecognition, I have decided to look at whiteness, to stare it down to its parts, to see how it works and name it. My focus fell on white literature.

In its broadest sense, white literature is not only the literature produced by white authors – individuals who occupy and perform the structural position of whiteness, but any text that exhibits, operates on the level of, and works for whiteness. Just as white people are "not represented to themselves as white," so is white literature rarely self-reflexive of its whiteness, and not labelled such within dominant discourse. Jess Row explains that white writing

normalizes conditions of language restriction that would otherwise appear dysfunctional, even diagnosable; but it does this without describing those conditions as problematic ... White writing is a covenant, a shared understanding, about what is sayable and what is shameful and not allowed. (92–93).

In selecting the texts for the present book, this characteristic of whiteness became very clear to me. None of the writers I encountered were canonized as white although they 1) were and wrote from the position of being white, 2) built their stories around white protagonists, and 3) openly juxtaposed these protagonists to racialized-as-Black secondary characters, i.e. activated the divisions and workings of racialization. Instead, the writers I was researching were understood through their
- gender and sexuality as in the case of Kathy Acker whose being a queer feminist not only framed but became the most important feature of her oeuvre (see Kraus, Pitchford)
- political and religious orientation as we can see in Marilynne Robinson's reception as a religious writer thanks to her recurrent thematizations of Calvinist thought (see Engebretson)
- or, were understood as *just human* as we can see in readings of Don DeLillo as a writer who has allegedly surpassed individual difference (like race and locality) and is therefore read as "plural," "essential," and universal (Row 86).

By pushing forward the whiteness of the texts I examine, by rendering it a point of focalization and a major characteristic of these texts, I depart from the above-mentioned interpretations. Instead, I join a growing number of literary critics and writers who have recognized the necessity to study the racialization of writing and reading practices – Toni Morrison, Hortense Spillers, Christina Sharpe, Sabine Broeck, Jess Row, and others. Thus, I take on the possibility to re-think white texts, moving away from the variety of what I would call *surface* (performed; aesthetic as opposed to structural) aspects. Instead, I propose to analyse the synergies, overlaps, and repetitions between novels like Kathy Acker's *Kathy Goes to Haiti*, Don DeLillo's *Zero K*, and Marilynne Robinson's *Gilead* as stemming from, and afforded by, the structural position of whiteness. My study is, therefore, equally focused on

1) glitches and lapsus when whiteness's assumed/unspoken grammar betrays itself (Wilderson, *Red, White, and Black* 5), i.e., performances of blunt whiteness and

2) on whiteness's buried pulls*pulse, i.e., the workings of whiteness without and beyond visible performance.

The texts and writers I discuss have little in common if we consider their writing styles, thematic interests, and audience. In fact, to claim that they belong to the same canon might sound quite preposterous. What could a punk icon of experimental literature like Kathy Acker have to do with Marilynne Robinson's soothing, beautiful prose? How can they relate to Don DeLillo whose claustrophobic narrative worlds resemble neither Acker's rowdy feminism, nor Robinson's devout optimism for humankind? How could *Kathy Goes to Haiti, Zero K*, and *Gilead* join and repeat the same story? None of the existing differences between these writers and their texts seem irrecuperable if we examine them through the perspectives of Afropessimism, Critical Race Theory, and Critical Whiteness Studies. On the contrary, the proposed symptomatic readings show how an interconnection emerges, and other aspects of the texts become visible, diagnosable.

I zoom in on one of these aspects – avant-gardism – and examine whether and how it appears in and as white literature. To do this, I follow two main routes: first, I analyse whether the selected material promotes such avant-gardism, for example in the narrativization of renewal or by articulating cutting-edge ideas; and second, whether the selected material itself is read as avant-gardist by literary critics. In both these questions, I consider avant-gardism's relationship to whiteness. Thus, I look at how avant-gardism operates in and outside of the texts – whether and how they push renewal forward with and as white authority/authorship. Lead by the interest in whiteness as authority that seeps in and out of texts, I sometimes conflate narrators, protagonists, and authors.[4] This conflation allows me to expose and highlight whiteness (and one of its core tenets – avant-gardism) and how it

---

4 I mark the conflation by using "author/ity" instead of authority; the conflation goes against the frequent divorce between racist fiction and its producers (consider debates about racism in texts by canonical figures like Mark Twain, Dr. Seuss, Roald Dahl, etc.)

works for white figures (independent of whether they are narrators, protagonists, or authors). It should be noted that the conflation overlooks the differences between real and fictional personages for the more targeted discussion of whiteness.

Here, I allow myself the sweeping statement that, in a certain way, all white writing is avant- gardist, or that one basic feature of whiteness is the claim about renewal. If we consider the whiteness of the avant-garde canon, this statement does not sound sweeping at all. Critics have already pointed to the racist exclusion and tokenization of non-white practitioners in avant- garde collections and collectives (see Hong, Winkiel, Sell). However, I want to step away from the critique of racialized canonization. Instead, I suggest that all white writing carries the ideological fabric of being avant-gardist because whiteness claims itself to be and organizes as avant-gardist. In other words, avant-gardism is a feature of whiteness and this feature performs in white writing even if technically this writing does not fall within the established avant-garde period and canon.

To corroborate this, we need two things – first, a definition of avant-gardism, and second, an analysis of whether and how this avant-gardism can be considered intrinsic to whiteness. Let me start with avant-gardism, or the capacity to be avant-garde. The *Oxford English Dictionary* defines "avant-garde" as "the foremost part of an army; the vanguard," and as "the pioneers or innovators in any art in a particular period." The *Cambridge Dictionary* adds to this definition "the painters, writers, musicians, and other artists whose ideas, styles, and methods are very original or modern in comparison to the period in which they live." In *Merriam Webster Dictionary*, "avant-garde" is explained as "an intelligentsia that develops new or experimental concepts especially in the arts." What the definitions have in common is the idea of abled, active bodies ("army," "innovators," "pioneers," "painters...," "intelligentsia") who advance or create something new. This is also how the term[5] came to be:

> French for "advanced guard," originally used to denote the vanguard of an army and first applied to art in France in the early 19th century. In reference to art, the term means any artist, movement, or artwork that breaks with precedent and is regarded as innovative and boundaries-pushing. Because of its radical nature and the fact that it challenges existing ideas, processes, and forms, avant-garde art has often been met with resistance and controversy. (MoMa)

---

5 in its current meaning, the term has been traced back to Olinde Rodrigues, (known also as Claude Henri de Saint-Simon), and his 1825 essay "L'artiste, le savant et l'induestriel" in *Littéraires philosophiques et industrielles.* Rodrigues states that art is the fastest way to social and political transformation and therefore artists should lead the way forward.

Beyond this broad definition of avant-garde, there is a rich discourse around its periodicity, formal aspects, representatives, and significance. Among the most notable voices in this discourse are Renato Poggioli (1968), Peter Buerger (1984), Benjamin H. D. Buchloh (2000), Clement Greenberg (1939), Paul Mann (1991), Harold Rosenberg (1994) and *other white men*. Several members of the Frankfurt School have also picked up the topic and wrote extensively about the connection between new art and social transformation – for instance, Max Horkheimer and Theodor Adorno in *The Culture Industry: Enlightenment as Mass- Deception* (1982) and Walter Benjamin in *The Work of Art in the Age of Mechanical Reproduction* (1935). Worth mentioning are also recent interventions such as Mike Sell's 2011 book *The Avantgarde. Race, Religion, War* and John Roberts' *Revolutionary Time and the Avant-Garde* (2015) both of which consider the term in relation to contemporary politics (but remain faithful to traditional conceptualizations of the avant-garde).

Feminist critics have responded strongly to the male-dominated focus and discourse around avant-garde studies. Among them are Susan Rubin Suleiman (1990), Rosalind Krauss (1985), Diana Crane (1987), Peggy Phelan (1993), Kristine Stiles (2000), Elin Diamond (1996), as well as a younger generation of feminists such as Jill Richards (2020) and Lauren Rabinovitz (2003). Recent criticism has also shifted to include a wider range of avant-gardes – instructive in this regard are the works of Kimberly Benston (2000), Elizabeth Harney (2018), Andrea Khalil (2003), Phillis Taoua (2002), and Amanda Stansell (2003). Thanks to feminist scholarship, we now consider the interrelationships between exclusionary, restrictive understandings of "avant-garde" and the violent omissions in the construction of its canon.

In *A Rightful Inheritance: Locating the Black Avant-garde*, Laura Winton argues that despite the predominantly white canon, avant-garde practice "inherently belongs to people of color ... because of the issues of marginalization ... and the need to push against normative forms of art and communication" (3). Although I agree with Winton, the argument I push herewith might sound contradictory to hers – at least before I make two important distinctions. Winton is right that the struggle for survival, which is how Black people are forced to exist in a white- dominated world, can be called avant-gardist. For, the struggle is premised on experiment and resistance, is generally pushed to the margins, and dependent on Black people's courage to move a step ahead and into dangerous zones (which is what the term "avant-garde" first came to signify). Similar arguments have been made about the capacity of avant-garde practices such as jazz to best illustrate Black experience (see Okiji, Moten).

Unlike Winton, I use avant-gardism in a more stylized manner – in the meaning of recourse towards renewal as suggested by Buck-Morss. Whereas Winton works with different spectra of avant-gardism (marginalization, experimentalism, resistance, newness, etc.), I have mainly used the term as a proxy for the property to renew/the property of renewal. It is worth mentioning that many of the aspects of avant-gardism Winton discusses could be studied in the novels I selected for this book. I have refrained from such analysis, and instead concentrate on renewal and whether and how it relates to whiteness. From this perspective, saying that I will re-read white avant-garde literature sounds like a tautology – for there is an avant-gardism inherent to and claimed by whiteness.

In fact, the very definition of whiteness entails a recourse towards renewal. Let us look closer at the term's emergence in the North American context[6]. In its present meaning and use, whiteness first occurred toward the end of the seventeenth century – when the ruling class effected the superimposition of class consciousness with race consciousness by instituting privileges for belonging to "a white race" and suppressing the rights of those who were not part of it (Th. Allen 239–259). As Theodore Allen explains, this move effectually cancelled labour solidarities between indentured European servants and enslaved Blacks in America as they had manifested during Bacon's Rebellion (1676–1677). For poor European servants, writes Allen, upward social mobility was a counterfeit concept – they had to be

> satisfied simply with the presumption of liberty, the birthright of the poorest person in England; and with the right of adult males who owned sufficient property to vote for candidates for office who were almost invariably owners of bond-laborers. (248)

To keep its social and political control, the ruling class needed to prevent insurrection and the possible union between indentured white servants (whose labour conditions were not too different from those experienced under slavery) and enslaved Africans. Allen writes:

> The solution was to establish a new birthright not only for Anglos but for every Euro- American, the "white" identity that "set them at a distance" ... from the laboring-class African-Americans, and enlisted them as active, or at least passive, supporters of lifetime bondage of African-Americans. (248)

---

6 For the purposes of my analysis, I focus explicitly on the antagonism whiteness – Blackness in the North American context. I also use definitions of whiteness and Blackness that emerged in radical Black studies in North America, and more specifically, Afropessimism. It should be noted, however, that whiteness and Blackness have been conceptualized in different settings – see for instance the work of Steve Biko and Édouard Glissant who wrote in and about the South African and the Caribbean contexts respectively.

Whiteness becomes meaningful in this moment of activating privileges. Those positioned as white were immediately promoted to a higher social class independent of their working or living conditions (Allen 290). In fact, several decrees were issued to improve these conditions. Allen notes that two types of policies were implemented around that time – policies that increased white people's rights (e. g. outlawing whipping of white servants, setting a minimum wage for them, allowing them to own livestock, etc.) and policies that explicitly targeted free Black and Indigenous people by denying them the right to be active, albeit poor, participants in society:

> Such were the laws ... making free Negro women tithable; forbidding non-Europeans, though baptized Christians, to be owners of "christian," that is, European, bond- laborers; denying free African-Americans the right to hold any office of public trust; barring any Negro from being a witness in any case against a "white" person; making any free Negro subject to thirty lashes at the public whipping post for "lift[ing] his or her hand" against any European-American, (thus to a major extent denying Negroes the elementary right of self-defence); excluding free African-Americans from the armed militia; and forbidding free African-Americans from possessing "any gun, powder, shot, or any club, or any other weapon whatsoever, offensive or defensive." (250)

These laws jumpstarted the long durée of white ideology – the form of consciousness that positions white people as legitimately superior, active, and abled contributors to and participants in society. The avant-garde of this invention, the (re) new(ed) status quo that aimed to preserve the dominance of the ruling class, and to continue the oppression of Black and Indigenous people, consisted in the promise of a better life that could be generated from white positionality:

> Thus was the "white race" invented as the social control formation whose distinguishing characteristic was not the participation of the slaveholding class, nor even of other elements of the propertied classes ... What distinguished this system of social control, what made it "the white race," was the participation of the laboring classes: non-slaveholders, self-employed small-holders, tenants, and laborers. (251)

Whiteness entailed that lower white classes were better than enslaved Black and Indigenous people but also, and quintessentially so, that their working and living conditions were subject to improvement and progress. For, the relationship occurring between policy and whiteness was both symbiotic and dynamic – policy began to authorize whiteness and whiteness to authorize policies in return (see Lowe, Hartman, Lopez). This made whiteness self-sustainable in principle; a synonym for an inborn activity and ability that can bring white people forward (some, like the ruling class, literally, and others, like poor white Europeans, potentially). In this sense, whiteness became a promising and a positive category – one connected to the (well)being and future prospects of those who inhabit it. Recalling white authorities from Herodotus to Thomas Carlyle, in 1851 John Campbell infers:

> The white never loses, but always gains. A nation or a tribe of the white race may become extinct, from a variety of causes, but the civilization of the race progresses notwithstanding … It is the nature of the white to progress, this appears to be the fact from all history, from all experience, from all nature. (*Negro-Mania*)

If we look closer at white ideology, we will see just how crucial to its functionality is such promise for a brighter future. Starting off as a set of privileges that implied better living and working prospects, whiteness became synonymous with that promise – even if the promise did not pan out for everybody as Allen shows (248–270), and even if the meaning of "a better life" changes over time.

We see the hold of this promise in white characters like Kathy in Acker's *Kathy Goes to Haiti*, Jeffrey in Don DeLillo's *Zero K*, and Jack in Marilynne Robinson's *Gilead*. Independent of what is their plight or path in life, independent of plot twists and creative distances between texts and techniques, all these characters use the authority of whiteness to claim and reach a brighter future. What is more, their authors are themselves praised for marching towards it – as my analysis reveals, sometimes against plain evidence to the contrary. This owes to an avant-gardism embedded in whiteness – one that positions white figures as active and abled bodies always already in movement towards a better tomorrow.

At the other end of this promise lies the construction of Blackness as the opposite of activity and ability (and their derivatives – progress, movement, rebirth, transformation, transcendence, etc). White colonizers, writes Franz Fanon, framed the Black as

> impervious to ethics, representing not only the absence of values but also the negation of values … the enemy of values … absolute evil. A corrosive element, destroying everything within his reach, a corrupting element, distorting everything which involves aesthetics and morals, an agent of malevolent powers, an unconscious and incurable instrument of blind forces. (*The Wretched* 6)

As blind force, Black people appear not only as a threat upon themselves but upon the functionality of the entire social and political organism – the "civilization" in Campbell's words. Thus, Blackness was rendered dangerous to the very potential of historical progress and human existence (as Campbell underscores, this existence is premised on white renewal, i.e. it is sustainable even when some white people "become extinct"). As a signifier for all that is positive and possible in and as life (being, creation, presence), whiteness juxtaposes Blackness that embodies its counterpart: negativity, stasis, and death.

"If the Black is death personified," writes Wilderson, "the White is the person-ification of diversity, of life itself" (*Red, White, and Black* 43). Crucially, this life is generated in the antagonism, through the violence, between whiteness and Black-ness. It is not simply that whiteness stands for positive categories – whiteness sig-nifies in the process of closing these categories to Blackness, when it finds them absent in and antithetical to Black being. Or, white and Black positionality are "not only opposed, but dialectically so ... the one positive, the other negative, de-pends on the other" (Wynter, "Towards the Sociogenic" 40). As Wilderson points out, whiteness emerges and operates in and as this antagonism (*Red, White, and Black*). To stay meaningful, whiteness depends on maintaining the process, i.e. maintaining the violence of negating Black capacities and reaffirming white ca-pacities. Steve Martinot and Jared Sexton explain:

> White supremacy is nothing more than what we perceive of it; there is nothing beyond it to give it legitimacy, nothing beneath it nor outside it to give it justification. The structure of its banality is the surface on which it operates. Whatever mythic content it pretends to claim is a priori empty. Its secret is that it has no depth. There is no dark corner that, once brought to the light of reason, will unravel its system ... its truth lies in the rituals that sustain its circu-itous, contentless logic; it is, in fact, nothing but its very practices. (175)

Whiteness *is* the continuous repetition of anti-Black practices, in a way that white people "recompose and reenact their horrors on each succeeding generation of Blacks" (Wilderson, *Red, White, and Black* 55). In this instance, repetition is nothing other than renewal – the continuous re- birth and re-appearance of whiteness, of white being. From this perspective, whiteness simultaneously signifies and func-tions as avant-gardist, i.e. it owns renewal as property (as its possession *and* its characteristic). As Wilderson writes, "anti-Blackness is the genome of [the] tem-plate for Human renewal" (337).

Because whiteness emerged from and as authority, anti-Black practices be-came the order of the day – they became normalized, legal, and coherent with so-cial progress and morality (see Du Bois, Morrison). Within the logic of whiteness, Black bodies appeared as negative (dangerous, inferior, backward, static) and as nothing (Non-Being, Non-Human, dead). Calvin Warren explains:

> The world needed a being that would bear the unbearable and live the unliveable; a being that would exist within the interstice of death and life and straddle Nothing and Infinity. The being invented to embody black as nothing is the Negro. An anti-Black world desires to obliterate black as nothing – nothing as the limitation of its dominance – so that its schematization, cal-culation, and scientific practices are met unchecked by this terrifying hole, nothing. With the Negro, metaphysics can triumph over this nothing by imposing black(ness) onto the Negro and destroying the Negro. The Negro is invented precisely to be destroyed – the delusion of meta-

physics is that it will overcome nothing through its destruction and hatred of the Negro. The Negro, then, is both necessary and despised. (*Ontological* 37–38)

Taking Warren's discussion into account, we can see how the avant-gardism of whiteness, whiteness' being, is structurally contingent on the destruction of Black bodies. Etymologically speaking, this interdependence reflects the first uses of avant-garde – avant-garde groups advanced and came to be but did so in a battlefield, by destroying an enemy, by cancelling something unwanted.[7] From this perspective, saying that whiteness operates as avant-gardist does not contradict Winton's idea of avant-gardism as a quintessentially Black practice. For, marginalization, experiment and struggle (one side of avant-gardism) are directly necessitated by advance and (re)birth (the other side of avant-gardism). In other words, renewal and destruction are fused: the moment whiteness advances forward is the same moment Blackness is destroyed; they happen on the same field, at the same time.

Following Hortense Spillers and Saidiya Hartman, Warren underscores that whiteness instrumentalizes Blackness – renders it an equipment, a material – for that very reason (*Ontological* 44–48). In other words, whiteness summons Blackness to cancel it out, for its unremitting cancelation re-affirms white being. Wilderson writes:

> [Black people] are being genocided, but genocided *and* regenerated, because the *spectacle* of Black death is essential to the mental health of the world – [Black people] can't be wiped out completely, because [their] deaths must be repeated, *visually* (*Afropessimism* 225)

This argument clarifies how the positive, the often celebrated, side of white avant-gardism (renewal, progress, re/invention) not only cannot exist without, but also generates the other side of the coin – Black destruction.

Thus, my reading turned into an investigation for those moments in which positive categories such as renewal, creation, and futurity reword the violence they rest on, i.e. blunt and bury the production of Black negativity and acts of Black negation embedded in these categories. In this regard, the present book follows a simple itinerary – I analyse the portrayal of white avant- gardism in light of the destruction of Blackness, pausing at three crucial stops on the way to this portrayal.

---

7 later this idea was taken metaphorically – unwanted became old traditions and the past as the time opposite of that what was not yet there. In its essence, the avant-garde is always already in opposition to what it does not desire

Movement

Firstly, I examine renewal by looking at the movements through which white figures achieve it. In its essence, renewal contains the distance between a current and a future state of being. This distance needs to be "gone" for renewal to be there: the old being is left behind/aside while the new being is reached/returned to. It should be noted here that even if renewal implies a return to a previous time, i. e. if it is achieved through recurrence or repetition rather than the creation of something original/new, an alteration transpires that marks the transition from state A to state B (Ladner 13–15). In fact, no matter what the model of renewal is, one requires ability and activity to bridge the distance between the first and the second, renewed, state. Nothing can be renewed if it stays the same, in one place, static.

In the American context, this idea holds particularly true. It is, in fact, fundamental to the construction of Americanness (see Adams, Bercovitch, Jillson). "It is something strictly American," writes Gertrude Stein, "to conceive a space that is filled with moving, a space of time that is filled always filled with moving" (226). Crucially, the movement Stein refers to is neither chaotic, nor ineffectual. As Sacvan Bercovitch explains in *The American Jeremiad*, one of America's constitutive myths is the idea that the American nation is headed towards a promising future, that its transformative potential and capacity to progress brings America to an ever-expanding horizon of growth, possibility, and renewal.

Within this mythology, America appears "as an unfolding prophecy," always already on the move to its realization (xiii). The actions and abilities of those who participate in Americanness, their individual and collective advance forward, bring about the prophecy's fulfilment. Rather than a halt at some final destination, however, the fulfilment transpires as the sustainability of process/progress, a continuity of (social) movements that keep America alive. For albeit changing, America does not distance itself from its own image because the capacity to change, to move from state A to state B(etter), is entrenched in this image. Bercovitch notes:

> "America" … is a powerfully composite, extraordinarily flexible cultural system sustained to a remarkable degree by the authority of dissent … The United States has insisted on renewal and change – or more precisely, as Barack Obama's 2008 slogan put it … "change we can believe in": forms of renewal that confirm the basic tenets of the system. (xxvi)

Dissent, or opposition aiming to transform the system, materializes both literally and metaphorically as movement – just think how movement is instrumentalized when protesters march, block streets, and do sit-ins, or how movement marks the exchange and spreading of ideas, and shifts of consciousness. In fact, dissent could not happen if people were not moved (personally and politically), i. e. if they did not feel socially engaged and prompted to act. Neither could it happen without people's ability to self-organize, to go against the system, and demand change. To move and be moved lies at the heart of transformation, and as Bercovitch shows, at the heart of America's reinvention (93).

The idea that movement(s) can bring about a better state is thus not only a part of the stories I analyse but their background, the context in which these stories unfold. Authors like Kathy Acker and Don DeLillo have been praised for moving past aesthetic hegemonies and traditions, and not merely because of their formal experimentations and originality. Both Acker and DeLillo have responded to the urgencies of their times, writing about and against what threatens freedom in its specific American realizations. Rebellious and politically conscious, their writing continues to be seen as transformative potential, a visionary contribution to America's progress and an inclusive, *woke* American-ness, i. e. a build-up to a positive, promising renewal.

I became interested in the ways Acker and DeLillo thematize this movement and the arrival at a better state. More specifically, how did they tell the story of white character(s) moving towards renewal? Was this a self-reflexive comment on their own positionality, an attempt to strengthen or withdraw from it? How far did their subversive rhetoric go, and did it confirm "the basic tenets of the system" as Bercovitch claims? If so, how was such transformative potential spared the criticism of being feigned or failing? Or does whiteness spin on a loop, always already appearing as new, future-orientated, and successful?

To investigate these questions, I focused on Don DeLillo's 2016 *Zero K* and Kathy Acker's 1978 *Kathy Goes to Haiti*. Although the texts can be deemed minor for the lack of serious critical attention,[8] both Acker and DeLillo are by now canonical – authorities in American literature. One of my concerns was to see to what extent reputation, as a manifestation of authority, plays a part in the working of and with these novels. Thus, I examined movement not only in *Zero K*'s and *Haiti*'s narrative worlds, but also in the ways these texts entered and were positioned within literary discourse. In all these questions, I consider whether and how white authority regulates movement within literary creation and interpretation.

The close readings lead me to the following wrap-up: whiteness operates on multiple levels – from the construction of white heroes and the way they progress textually, through the racialization of language and of reading practices, to the romanticization and rescue of white figures in popular discourse. In all these, whiteness's positive avant-gardism functions by and comes at the cost of ignoring how

---

8 perhaps due to *Zero K*'s recent publication and the niche role of *Haiti*

Blackness is negated and rendered negative. *Zero K* and *Kathy Goes to Haiti* demonstrate how white author/ity uses this negativity to (re)invent itself, how it moves in and out of texts as an imaginative capacity, a regulation of perspective that always already shifts the lens from the destruction of Blackness to the renewal of whiteness.

Futurity

Having discussed the build-up to renewal, or how white figures advance towards it, I consider renewal's representation. Both *Zero K* and *Haiti* position renewal in the future of white characters, hence a section dedicated to white conceptualizations of futurity. In this section, I analyse what John Dos Passos once called the "great white curtains of eternity" (240), the horizon of progressive history white protagonists and white literature see as a promise and a destiny. Here, eternity signifies the time in and as which renewal takes place, the time reached by white characters' successful movement in history (Bercovitch 45–51). My main concern is whether and how this time is racialized, i.e. whether or not whiteness regulates images and imagination of this time.

To begin with, let me briefly outline how I use the term futurity. In the *Oxford English Dictionary*, the term is explained as

> the quality, state, or fact of being future; future time, a future space of time; what is future. that is to be, or will be, hereafter; ff or pertaining to time to come describing an event yet to happen; a condition in time to come different (esp. in a favourable sense) from the present.

Of particular interest to me is the idea that futurity signifies "a different" time to come, or that a change has taken place that separates the future from the past and the present. As we have seen from the previous section, white author/ity (in the examples I give – manifested by white literature) is said to carry the potential of generating this transformed time. We can see this idea unfold in the context of creative activity. As Amir Eshel writes, futurity

> marks literature's ability to raise, via engagement with the past, political and ethical dilemmas crucial to the human future. In turning to the past, the works ... keep open the prospect of a better tomorrow. Many ponder the human capacity to face hopeless individual or sociopolitical circumstances. Yet precisely by engaging such circumstances, they point to what may prevent our world from closing in on us. (5)

For Eshel, literary engagement with the past, and more precisely, with past social and political crises, opens the possibility to create a better future. Similarly, Sacvan Bercovitch reminds us of

> Vladimir Nabokov [who] recounts the story of an ape that, "coaxed by a scientist…produced the first drawing ever charcoaled by an animal: the sketch showed the bars of the poor creature's cage." A depressing outcome in Nabokov's telling, but by analogy to human animals, who build their own cages, *that sketch [is] the indispensable first step in opening vistas of political transformation* (xxiii, emphasis mine)[9]

We will see hope for political transformation play out when Acker appears as a rebel and DeLillo as a savior of history thanks to their narrative choices to name oppression. By articulating the existence of human failures and crises, whether they are criticized and deconstructed or not, Acker and DeLillo emerge as *seeing*, and capacitated to produce alternative visions – i.e. they appear as visionary. Eshel writes further that in the context of imaginative and creative ability/activity futurity designates "the potential of literature to widen the language and to expand the pool of idioms we employ in making sense of what has occurred while imagining whom we may become" (5).

I was interested to see whether subversive white writing can be severed from past, traditional burdens of whiteness. Can representation be free of the authority that creates it and the world it enters – where this authority dominates? If whiteness seeps into images and language, if it regulates the lens, then how does white literature conceal its constraints and limitations and instead celebrates its freedoms, open potentiality, and generative newness? I have posed this question earlier – how does white avant-gardism obscure anti-Black destruction when both are right there, at the same spot? What adjustment of lenses would be needed for a conscious, more critical, capture of both sides of renewal when we consider that whiteness always already alibis itself out? Is an adjustment possible when racialization rests on the retina as Dionne Brand says, and even imagining the future, the better time – is not safe from it (Sharpe "Response")?

---

9 I focus on Bercovitch later (see *Newness*). I want to pause here and mention that one important reason I re-read his work becomes evident in this quotation – namely, the problematic way in which Bercovitch subsumes everybody in the category "human animals, who build their own cages." Historically, this is not the case. One does not even need a metaphorical **[jumpcut]** to think of chattel slavery, the prison industrial complex, or quite bluntly: of white people entertained at Human Zoos, to recognize the incommensurability between the positionalities between those people who built cages and people who were locked in them (see, for instance, Blanchard's *Human Zoos*).

Eshel is not the first to tie futurity to the capacity for ethical and political transformation, the better tomorrow where things begin anew. As Hannah Arendt writes, "every end in history necessarily contains a new beginning" (478–479). In fact, this coda can be traced back to the earliest philosophical theorizations of human existence and history (North, *Novelty* 22–36). What interests me is the trajectory between horrifying past – better future which seems integral to the capacity to be (at least as far as western configurations of being are concerned). This coda for ever-renewable being, its sustainable futures, however, looks very different if considered alongside Black critique.

In fact, working with and through the past does not always guarantee recuperation – at least, not in the sense of the past being finally undone, done with (see Wilderson, Sharpe, Sexton). For instance, when Tina Campt "goes back" to re-think old Black photographs in *Listening to Images*, she does so with rhetorical scepticism: "what does it mean for a black feminist to think about, consider or concede to the concept of futurity" ("Black Feminist Futures")? This is a difficult question, and one that has arisen against celebrations of progressive history and alongside the theorization of an alternative view of time, the "always now" of anti-Blackness, of white authority at work (Morrison 210). This time locks, or holds (to use Wilderson's term), everyday Black existence in the repetition of terror (remember, the white world is and continues to be through the repetitions of anti-Black practices as Martinot and Sexton say). Christina Sharpe calls this repetitiveness an existence in the wake:

> Living in the wake means living the history and present of terror, from slavery to the present, as the ground of … everyday Black existence; living the historically and geographically dis/continuous but always present and endlessly reinvigorated brutality in, and on, [Black] bodies while even as that terror is visited on [Black] bodies the realities of that terror are erased. (15)

Sharpe is not alone in underscoring the endless dis/continuity of anti-Black terror. From the perspective of Afropessimism, the negation of Black being did not end or expire. On the contrary – it is always-happening, a temporality in which past, present, and future are equally superimposed (horrors haunting) and stricken (free life unlived) (see Wilderson, Sexton). In other words, the freedom of a new and better time does not lie at the horizon of this time's renewability, but as Wilderson shows pace Fanon – at the end of the world, at the end of whiteness progressing:

> For the Black, freedom is an ontological, rather than experiential, question. There is no philosophically credible way to attach an experiential, a contingent, rider onto the notion of freedom when one considers the Black – such as freedom from gender or economic oppression, the kind of riders rightfully placed on the non-Black when thinking freedom *[recall Acker and DeLillo here]*. Rather, the riders that one could place on Black freedom would be hyperbolic –

> though no less true – and ultimately untenable: freedom from the world, freedom from Humanity, freedom from everyone (including one's Black self). (*Red, White, and Black* 23)

What Eshel forgets in his definition of futurity, and what becomes clear from interventions like Campt's and the Afropessimist demand for the end of the world, is that the capacity to move forward and reach a (liveable) future, i.e. the propertization of Phoenix itineraries (from utter destruction to social betterment), is a capacity claimed by some and denied to others. Not every destruction and horrific past can be paused and worked through by subsequent generations, not if the destruction is practiced ad infinitum and the continuance of the practice fuels the functionality (or what Wilderson calls *psychic health*) of the world (*Afropessimism* 329).

In the hold of anti-Blackness, returning to the past cannot disinter a better tomorrow from the ruins – a point made painfully clear in Saidiya Hartman's *Venus in Two Acts* and M. NourbeSe Philip's *Zong!* As Hartman and Philip show, the destruction of Black bodies is not only still ongoing but obscured by the available modalities of memorialization. Closure is impossible because the modes in which it functions are insufficient and flawed, and because the destruction of Black being is everywhere, always, "the continuous and changing present of slavery's as yet unresolved unfolding" (Sharpe 14). "One can't even begin to think about memorialization," writes Hartman, "because people are still living the dire effects of the disaster" ("Memoirs of Return" 111). Philip notes,

> I deeply distrust this tool I work with – language. It is a distrust rooted in certain historical events that are all of a piece with the events that took place on the Zong. The language in which those events took place promulgated the non-being of African peoples, and I distrust its order, which hides disorder; its logic hiding the illogic and its irrationality, which is simultaneously irrational. (197)

Because memorialization is both complicated and implicated, "the story that cannot be told must not-tell itself in a language already contaminated, possibly irrevocably and fatally" (199). For Philip, there is neither retrieval, nor redemption – even if one goes back to the stories that

> mark the moment of destruction. These stories are *visually* open, "happening always – repeating always, the repetition becoming a haunting" (203).

Philip's resistance to a fatally contaminated language is to cut it open, break its teeth – its grammar, and let a sound be (203). As I have written elsewhere,[10] this

---

10 See Nikolova, *Of the Fugue in the Passage to Madness* 2–67.

sound remains incomprehensible to white audiences, beyond white time and white space, "at a lower frequency" (Gilroy 37), where "words don't go" (Moten 59). We can see this cut practiced in *Zong!* but also in Toni Morrison's *Beloved* as a number of other critics argue, and we can see it theorized as well. Christina Sharpe, Fred Moten, Saidiya Hartman, and Édouard Glissant are just a few of the many Black thinkers who have discussed and practiced "the cut" as refusal, as a critique against the limits imposed on Black being.

Could the cut be contained in, and spring from, Eshel's understanding of the "futural perspective," the endless number of possibilities and progress arising with the return to the past? This question is not for me to answer. I will say this, however – when Eshel thinks about "a futural perspective" as growing from past trauma, he premises futurity on "responsibility, ethical and political action, and notions of justice" (12). Yet, the very notions of justice, ethics, and political action are anti-Black (see Vargas, Hartman, Warren) and as Wilderson argues, "Black people *embody* … a meta-aporia for political thought and action" (*Afropessimism* 13). Is it possible to build a free Black future from the ruins and hopes of an anti-Black world? The Afropessimist answer is – of course not. Yet, white innovative writers like Acker and DeLillo are said to do precisely this. Disagreeing with Acker and De-Lillo's reception, my analysis joins Afropessimism[11] and highlights the divide between white and Black work – and the irreconcilable futures they lead to.

---

11 While my analysis is informed by and directly linked to Afropessimism as a critical framework and perspective, it also engages the work of Black scholars who do not consider themselves and are not considered Afropessimist (for instance, Saidiya Hartman, Christina Sharpe, Fred Moten, Édouard Glissant, Orlando Patterson). I do not assume that these scholars (and Afropessimists for that matter) agree on all points, nor that their work is reducible to a single theoretical whole, lacking original, unique, and distinct interventions. I do see shared/intermixed moments between and in these scholars' texts (in the sense of jazz' *sitting in*, or Sharpe's *sitting with*), however, and see these moments revealed through an Afropessimist lens. In my reading, one such moment is the radical refusal of the white world and white world-making, or anti-Blackness as Afropessimists would call it. Another such moment is the argument that anti-Blackness is pervasive, normative and operating in both obvious and invisible ways. I am little concerned with recasting scholars like Hartman, et al as Afropessimist (I write this project in the spirit of walking the same path respecting the distances between us, as discussed earlier). I also do not want to erase the criticism of and against Afropessimism (see, e.g., Asad Haider's *Mistaken Identity: Race and Class in the Age of Trump*, Lewis R. Gordon "Shifting the Geography of Reason in Black and Africana Studies", "Critical Exchange: Afro Pessimism", Greg Thomas' "Afro-Blue Notes: The Death of Afro-pessimism (2.0)?.", etc.) nor the difference this criticism sees between the above-mentioned scholars' and Wilderson's work (as a forefather of Afropessimism). I have left aside other possibilities and perspectives through which the selected corpus may be and has been examined: 1) because they are not the purview and concern of the present book, and 2) because my readings of Hartman, Sharpe, Moten, Glissant, and Patterson are linked to Afropessimism. My engagement with Black 'non' or 'anti' Afro-

On the one hand, whiteness is always already invested with futurity, the promise of its re- emergence. As Wilderson explains, whiteness is

> a structure whose idiom of power is autodidactic and autoproductive: it generates its lessons, its ensemble of questions and their attendant ethical dilemmas, and its institutional capacity, internally, without recourse to bodies or questions beyond its own gene pool ... Whiteness has an infinite ensemble of signified possibilities: The infinite possibilities themselves cannot be definitely named; their dramas of value cannot be predicted with anything approaching precision; nor can the reproduction of these possibilities be threatened with mortality, because Whiteness's internal mutation is limitless. (*Red, White, and Black* 297)

White futurity, not only white life, is liveable and positive – even if white people are mortal, futurity marks an always approaching, renewable presence (recall Campbell). According to Afropessimists like Wilderson, the future time for Black people is "the always now," the repetition of death and destruction inflicted on Black bodies – the hold of nullified past, present and future, sucked into one a-temporality (*Afropessimism* 216–19). There is a quintessential difference between the repetition of life – renewal – and the repetition of death – nothingness. One way we can capture and consider this difference is if we focus on how white writing frames future time for its white characters and future time for Black characters, and on how whiteness reacts to alternative portrayals, to Black demands for a liveable Black future.

What flashes in white writing, what becomes pronounced and visualized, is precisely this ideological mechanism – the unfolding of whiteness as moving towards renewal, white being repeated in progress, and the violence that forces Black bodies to keep on dying, deadly and deathly, always. Black critique has exposed this chokehold – Orlando Patterson talks about permanent social death (*Slavery*), Saidiya Hartman about the afterlife of slavery (*Scenes*), Toni Morrison about the always now and no-time (*Beloved*), Christina Sharpe about wake and residence time (*In the Wake*), Frank Wilderson about the condition in the hold (*Red, White, and Black*), Èdouard Glissant about piling time (*Poetics*), Ta-Nehisi Coates about robbed time (*Between*). At the heart and inter*mixing of all these conceptu-

---

pessimist works transpires at this juncture and in this light. References to Black works that do not fall into the categorization 'Afropessimist' should not be taken as the relabelling of these works. In my opinion, shared/intermixed moments, conversations, and sitting ins/withs can occur between different interlocutors, traditions and schools of thought, styles/forms of expression, rhythm and register (thinking of jazz in more than one way here). Taking these shared moments and arriving at Afropessimist re-readings of white narrativization owes to the character of this book – a personal polemical meditation *with* Afropessimism as discussed earlier – and does not contradict nor dispute the fact that texts are slippery and giving to other and different readings and interpretations.

alizations lies the recognition that anti-Blackness, the work of whiteness, ceases and collapses Black lived and liveable time and instead authorizes its repetition in nothingness.

As Sharpe says, white renewal and Black negation are beyond anything else "questions of temporality, the longue durée, the residence and hold time of the wake" (*In the Wake* 22). Who gets to inherit and beget time – the always new – and who is robbed of it – hence, the always now? What acts of resistance can mark the refusal of this robbery? Discussing Abderrahmane Sissako's film *Timbukti* and the Black character of Zabou, Sharpe tells us that "Zabou doesn't believe in time, at least not linear time." This disbelief, Sharpe explains, arises with the fact that Zabou "lives in trans*Atlantic time, in an oceanic time that does not pass, a time in which the past and present verge" (128). As we will see, there is a fundamental difference between Zabou's disbelief and Black time, and white hope and white time.

I consider this difference by analyzing Marilynne Robinson's 2004 novel *Gilead* and by relating Robinson's images of white futures with practices of severing futurity from Black bodies (of work) – in Robinson's text and in public discourse. Robinson is an important writer to consider here – she is widely celebrated for her beautiful, almost therapeutic prose that is said to work through our human hardships and failures and leave the world a better place, her readers uplifted (see Lear, Muhlestein, Petit). If Acker and DeLillo are canonized as transformative due to their rebellious and blunt engagement with dominant discourse, Robinson is seen as transformative for appeasing it, for narrating the kindest of human qualities with unsurpassed gentility. Not surprisingly, her novels are compared to sermons, soft-spoken and inspirational.

Gilead is seen in the same light by white critics– uplifting readers' spirit and hopes, bringing the better in us and in our existence to the fore (see Lear, Petit). Therefore, I consider the conceptualization of white futurity both in Robinson's text, and in Robinson's reception. As for Acker and DeLillo, whiteness works not only as a representational schema in the creation of stories and meaning but as a regulation of readerly perspectives. My interest is to see whether and how Robinson's beautiful work gets rid of the ugly side of its ideological material, and if yes, what effect such concealment has on the existing critique of white ideology. In other words, I want to capture the process in which whiteness is aestheticized, its tracks covered, and functionality put into practice.

I look at two occasions of white authority at work. Firstly, I consider Robinson's intertextual plays with Toni Morrison's *Beloved* and examine whether *Belo-*

ved's critical energy gets distorted by alterations of Morrison's text in *Gilead.* [12] I then consider the forced removal of Alisha B. Wormsley's art piece "THERE ARE BLACK PEOPLE IN THE FUTURE." For the lack of a better term, I analyse the dismissal of Morrison and Wormsley's communications under the label "rewording" and show how white author/ity nullifies Black critique through it. Under rewording I understand the act of supressing one meaning – in this case, Morrison's and Wormsley's critique against anti-Black futures – and of advancing another – in this case the positive side of the coin, whiteness as promising a better tomorrow. It should be noted that rewording is an insufficient term but helps me to keep in mind that white renewal and Black destruction are simultaneous, symbiotic processes. Every part of what is reworded remains there – only the logic, the order of things, changes and with this regulates what we see, what *we can believe in.*

Newness

After discussing the racialization of renewal on a narrative and textual level, I re-read two theoretical accounts concerned with the new – Michael North's *Novelty: A History of the New* and *The American Jeremiad* by Sacvan Bercovitch. With these texts, I distance my analysis from fictional narrativizations of renewal. Instead, my focus falls on the process of crafting and working with its definitions. Thus, I can bring forth the similarities between creative manoeuvres enabling the racialization of renewal and their scientific counterparts.

Analysing how the definition of renewal was influenced by narrative and textual choices such as omission and incorporation, I raise the issue of the restrictive, partial nature of its conceptualization. I thus connect writing and reading, or knowledge production, to modes of understanding. Is the way of telling a story telling on its own? Does it pre-determine perspective and focus? As Black critique shows, white authority regulates not only that what is already here but that what could and urges to be, like imagination, alternative vision, other tactics of being in and knowing the world (see Moten, Hartman, Wynter).

---

12 I follow John Murillo in a rather *"specific* reading [of *Beloved*] from what is and has been a primarily Afropessimistic framework" (95). I do so departing from the vast and necessarily different interpretations present and possible when one analyses the novel with another focus and a different theoretical repertoire. Like Murillo, I embrace this "alternative reading" in its Afropessimistic momentum for tackling questions regarding "Black time, space, and creation" and their antagonistic relation to white time, space, and renewal.

I can best explain my interest in definitions – in what is missing or excluded within them – with a package I received a few weeks ago (I assume a gift for completing the present study). It was a copy of Lauren Berlant and Kathleen Stewart's *The Hundreds*, a bookmark left on "Not Over Yet." The page reads:

> You take the factory with you when you leave the factory town, the tinny smell of defrosted chicken shivering in its final moment. Telephones, too, remind you that you used to be willing to tether to something, even to lift the receiver to hear the ex say that you're still a piece of shit. But it's not over yet. Everywhere you went there was love and other kinds of disposses-sion. Everywhere you went you had urges without plans and sometimes you made plans. You can look around where you're are sitting now and know that what's there isn't all of it. (135)

I am interested in defining practices because they are the product of work and this work organizes and is organized by authority. Yet, definitions linger long after their authors leave "the factory," the context in which authority ordered, operated things. How we conceptualize renewal remains, its "tinny smell" bringing up the factory when the factory is gone, and so are we – somewhere else, beyond that mo-ment – hearing the same thing, a telephone resounding history.

Yet, "what's there isn't all of it." It is impossible to extricate the factory from what you remember, to block thoughts springing from *where* you knew, or felt, be-fore. But what you knew, what you felt – was never all of it. Examining *Novelty* and *The American Jeremiad* is my attempt to consider what is absent from the fabrica-tion, the fabulation of renewal. By thematizing the new, North and Bercovitch (re) create its definitions. The nature of this labour positions North and Bercovitch here, now – still/sitting and retelling history, describing what is said and done, but also in the factory, in history – working (while echoing traditions of produc-tion).

So, I ask – is it possible to pause, to consider what is outside of Bercovitch and North's visual field? Can we look around at new things and situations remember-ing that perspectives, the way we look, has been influenced by previous practice, and that neither the practice nor the product contains the full story? On the con-trary, the story is everywhere, everything, at the same time– things that we miss or deliberately avoid seeing, or that are *impossible to grasp* within the factory's walls. In the section *Newness*, I take what North and Bercovitch (re)produce, and try to bring it together with voices coming from outside the factory – voices that are ex-cluded, misheard, silenced, but *there.*

It should be noted that while I focus on the politics of omission and the absen-ces it generates, I refrain from revising *Novelty* and *The American Jeremiad* by in-cluding what they exclude. Instead, I read these texts with what is not present in them but is nonetheless there, as tension.

Physics teaches us that all matter, even this invisible to the naked eye, has mass

Novelty obscures Black histories and highlights white history. By spelling the latter out, North makes it perceptible to the reading eye, lends it access to the canon and the classroom guaranteeing its presence as a starting point for further explorations of renewal. Silencing other histories, North obfuscates their contours and forces them outside the Readers' visual field. Beyond the omission, however, the massive actuality of Black bodies (of work) exists. This actuality is an inextricable part of the history North set out to write. *Novelty* is readable this way too – with what appears in it, what does not, and with the tension between these two as a spectral presence; as a live antagonism between presence and absence that constitutes its own being, there, unspoken.

Unlike *Novelty*, *The American Jeremiad* includes Blackness in the concept of Americanness (and historical development). Subsuming white and Black perspectives and positionalities, however, Bercovitch makes Black being disappear. Different in performance but similar in principle, both omission and incorporation empty (overwrite, erase, reword) Black meaning and content. While it is impossible to re-insert, to re-imagine that what is absent (see Hartman, Philip, McDougall), at the end of this study I consider the protocols and practices creating this absence. Which is why I return to theoretical accounts of renewal such as *Novelty* and *The American Jeremiad.*

With the return, the present book goes in circles. For, any inquiry into definitions and the conditions from which they arise opens the possibility for further rereadings. Could we go back to Acker, DeLillo and Robinson and continue up a different path? Without sealing off the questions I began with by way of concluding this study, I offer you to pause and walk with me. I will tell you of the braces which helped me re-think white literature and whiteness as a living organism, in motion.

Sometimes, I give these braces names, like *rewording* and *jumpcut.* Other times, I need to leave the text I am analysing to find my footing – in calibrated* passages, far away from white stories and speech. Outside, I listen to Black voices who contextualize the theoretical impasse I am confronting (at home). Throughout, I follow Black critique. As might be expected, I tire, stumble, and fall. My foremothers would say this is part of the journey. They would also say that we could never arrive at the peak carrying the same knowledge, the same secrets, but this hardly enfeebles the feminist resolve to rise together.

[jumpcut]

Setting out to tell the story of how Whiteness claimed the future, I stray away from academic protocols three times.

1) In the passages calibrated with "*," I practice a type of "[n]arrative restraint," or what Saidiya Hartman calls "the refusal to fill in the gaps and provide closure" ("Venus" 12). The historical record of transatlantic slavery, Hartman explains, buries the enslaved in stories of excess and violence. These stories are how the enslaved surface in discourse and a life outside them – free life – is impossible. To imagine the enslaved through the scene of violence is to suspend them in the repetition of this violence. Hartman resists the repetition with "narrative restraint" – the refusal to know and say everything – and by interrupting the present with what is not said and known but is still there, unspoken.

Like Hartman, M. NourbeSe Philip uses narrative restraint and interruption to tell an impossible story – what happened to the enslaved Africans onboard the Zong. "There is no telling this story," Philip writes, "it must be told" (189). Philip does not re-imagine the massacre of Black bodies in 1781 and bases her book on the few-pages verdict from the so-known *Zong* trial. In the book, Philip destroys the verdict's text and coherence, and works in their break. Her exits of language puncture the archive and let the dead take space outside and against readerly capacity to see them, imagined. As I have written elsewhere,[13] *Zong!* is impossible to be understood by white audiences and this is Philip's refusal: she makes white audiences restrain themselves, catch out their desire for reading ahead, to reach closure and complete understanding.

I do not think narrative restraint applies only to what can or cannot be known about the past. I have felt my (white)[14] gaze and desire to see refused many times: by Èdouard Glissant's "right for opacity" (209) and the *only Black comments allowed* of networks like *Afrofuturist Abolitionists of the Americas* and *Black Agenda Report* on social media, or in the works of Fred Moten, Alexis Pauline Gumbs, and Toni Morrison. From where I stand, the refusal implies that visceral knowledges cannot be made legible for audiences who have long profited from denying them, and whose visual field blocks alternative perspectives. Trying to understand in universalized terms has produced only further damage and limitation. Still, if one attempts re- reading white (hi)story "against the grain," as Hartman says (*Scenes* 10), Black work is the knowledge to rely on.

Is there a way to learn from this work outside theft and accumulation, and without reducing it to useful material? During a conference in Vancouver, I met somebody who insisted on leaving the university premises to teach their classes. They said they want their students in the open, outside of the usual narrative/

---

13 See *Of the Fugue in the Passage to Madness*
14 I see myself positioned as what Wilderson calls white people's junior partner: a Bulgarian (i.e., an Eastern European who grew up learning about Bulgarian oppression and domination by the Ottoman empire), migrant, woman.

space ("the Universitas is always a state/State strategy" as Moten and Harney say [32] and the person I met wanted *out*).

I have needed the same break from white speech[15] and academic certainty and found it in the calibrated passages. There, I stray away from interpretation and try to avoid conclusions. I treat the passages as someone else's moments of strength, and I accept that a great portion of what is shared in them – strength required for escaping white author/ity in the first place – will remain a secret to me. I continue walking on the same path respecting the distances between us.

2) There are places in the text where I reach conclusions through **[jumpcuts]**[16] – movements that allow me to connect two points in narration with a missing part of the story **[jumpcut]** Sikho, a South African friend, tells me of "ndiYakuBona."[17] She says that in Xhosa villages, when a child is sent out late to get something and is scared of the darkness, someone from home would shout "ndiYakuBona! NdiYaku-Bona!" – I see you! I see you! Yet, the night had fallen and the caller could not have really spotted the child. Sikho tells me that beyond consolation for the child's fear, "ndiYakuBona" means that one can see something beyond the perceived image. Despite obstructed vision, the caller sees because they see with something other than eyes. I think of white (hi)story as fallen upon us, like night. It consumes everything around us and conceals the counter-histories, the "insurgent, disruptive narratives that are marginalized and derailed before they ever gain a footing" (Hartman, "Venus" 13). Yet, someone calls and someone sees, and someone is out there doing the work. In *Against Dominant Visuality, the Cut as Critique**, I include a few instances of how Black thinkers challenge dominant vision in the spirit of "ndiYakuBona." Using **[jumpcuts]** between conclusions, I follow Sikho and the people who refuse what has befallen as the only route back (home).

As a cinematographic technique, the jump cut breaks the continuity of time by moving from one part of the story to another. The movement entails the elapsing of time that remains outside what we see; those who see follow the story even if something is missing. The jump cut does not erase what happens outside of

---

15 For a white* reader, white literature causes simultaneously a rejuvenation (or *jouissance* to go back to Wilderson) and paralysis; it is both self-engulfing and safe (*Afropessimism* 92). I have tried to escape this vertigo in and through the calibrated passages – passages in which I read Black texts; passages meant to snap me out of whiteness. While necessary, the calibrated* passages are bonded to my ever-external readings and built through attempts to avoid interpretation and analysis. In a book on and about whiteness, the passages are necessary breaks from white speech in all the impossibility of breaking out (to follow Afropessimism here and else)

16 Marked with **[jumpcut]**

17 See also Sikho Siyotula's forthcoming book *Visualising*. I am referencing with Siyotula's permission and am deeply indebted to her for allowing me to read early drafts of the book

sight, and one cannot talk of what is skipped as absence. The skipped remains a secret. Black critique shows that the normalized means for making sense of the world is claiming authority over the secret, and that this normalization is violent and powerful (see Glissant, Hartman). The **[jumpcuts]** are an attempt to disown this authority if ever so briefly, to continue up the mountain skipping where skipping is a way of keeping time ajar, outside the (c)lock of understanding.

3) I use italics font to mark personal comments. In the process of writing, anger and exhaustion showed up on the page, often as sarcastic notes to myself. Such passages are generally discounted from academic works and understood to undermine the text as unfinished or unscientific. In *Talking Back*, bell hooks links the policing of what, how, by whom, and when something is said to the structural preservation of authority. She addresses academic writing specifically:

> When I began writing my first book I thought it necessary to abide by the academic training that had taught me not to focus on the personal, to maintain a tone that was scholarly. However as my engagement with feminist thinking and practice progressed I began to interrogate the notion of this more neutral-sounding academic voice ... I realized the importance of cultivating a writing voice that would empower me to speak about issues in a more open, almost conversational manner. (x)

For the longest time, I attributed my struggle with academic English to a discrepancy between my Bulgarian "voice" (if Italians are known to gesticulate profusely with hands, so are Bulgarians with words) and Western academia. Reading bell hooks, I have come to realize that the difficulty arises with a much stronger disavowal of the way some people speak and the legitimacy of their voices. I keep the "italics" moments as a reminder to myself – beware of formats; re-read what you put away in the margins; remember that rules can be learnt but so can the resolve to break them.

Movement

1 The Spectrum of white violence & spectral blackness in Don DeLillo's *Zero K*

Overview

Don DeLillo's work is often framed as a visionary and ethical reflection on contemporary crises. Such reception redoubles an idea of white American farsightedness and morality that is already embedded in many of DeLillo's stories – most recently in his 2016 novel *Zero K*. The following subchapter challenges celebratory rhetoric surrounding this narrative. Firstly, I consider how literary criticism works to position DeLillo in the line of American writers who address and thus metaphorically dismantle social evils (i.e., as a writer whose work strives for and achieves renewal). I then examine how *Zero K* enables visions of heroic/positive renewal despite and due to narrative exclusions, a politics of exclusion, and the repeated legitimization of white ideology.

Drawing a connection between DeLillo's reception and that of *Zero K*, I investigate the key lighting and key moments in creating renewal's image and how this image derives from and upholds white ideology. Black critique teaches us that, strengthened by discourse, white images dominate vision and perform work not only within their confines, in this case – the pages of DeLillo's novel, but work as and join in an economy of anti-Black imagination (see Bell, Hall, Wilderson). I use the following close reading to consider how novels such as *Zero K* participate in this economy no matter how ironical or self-reflexive they might (claim to) be.

As a subchapter of *Movement*, the following analysis prepares the ground for a more detailed discussion of renewal and focuses on 1) the movements that lead to it (e.g., the representation of protagonists' journey towards a better, renewed state), and 2) the extratextual movements that readers perform in interpreting these representations and relating them to renewal. In this regard, I look at movement both in its literal and metaphoric sense – movement to me implies the "change of place or position" that fictional characters enact but also the more involuntary flow of ideas, the "mental impulse" (*OED*) that transpires in the process of reading these fictional characters and drawing subsequent conclusions. Thought this way, movement points to the ways ideas and ideologies come in, or leave, the text.

Note: An earlier version of this chapter appeared in the 2019 Special Issue of COPAS, "White Supremacy in the United States and Beyond." Drafts of the chapter were also presented at a Symposium in Potsdam in June 2019 and at HKW (Berlin) in December 2018

https://doi.org/10.1515/9783110799996-002

This flow of ideas is also my main interest and concern in this section. I want to understand how white conceptualizations of the impetus for renewal work in and outside the text; how white imagination portrays this impetus; and what steps it envisions and takes towards renewal. As I will do in the subchapter on Kathy Acker, here I highlight a particular idea or image and examine its narrative function, some of its interpretations in literary criticism, and its relation to and role in dominant discourse. To zoom in on the idea/image, I have left the difference between text and author/ship, often conflating their spaces and searching for the work the idea/image does for them. This strategic manoeuvre results in a rather hostile reception of Acker and DeLillo (and the rest of the white writers I discuss).

However, this is not an intervention in the strict sense of #cancelDeLillo or #cancelAcker. Rather, I am interested to see how whiteness moves in and out of texts, how white writing promotes and defends its transgressions towards the racially othered. To follow Black critique, whiteness not only runs deep in obvious anti-Black practices but also sneaks in white subversive gestures and desires such as self-irony, self-reflexivity, and allyship (see Sharpe, Terrefe). I point my lens towards this process of "sneaking in," and, therefore, use DeLillo and Acker as material within which we can glimpse at whiteness moving. Undoubtfully, this use carries through both intentional and accidental hostilities towards Acker and DeLillo. The bracket I want to open here is that I consider DeLillo and Acker's novels an example in which I can study the movements of white ideas.

Therefore, I approach DeLillo and Acker's texts with a sense of shared responsibility for the mis/recognition and de/construction of whiteness writers and readers perform, united in and influenced by it. Which is why you will encounter a collapse of distance between fictional characters, authors, and me as a reader. This collapse is intentional and informed by knowledge articulated within Black critique – there is no ultimate absolution of whiteness for white people, no whiteness-free place where we can deny the occasions whiteness works for (see Wilderson, Sexton). What white people and their junior partners can do, however, is to make themselves literate in the ways whiteness operates, and to work against them. This is how I understand my own intervention and the purpose of the following analysis – how does whiteness come in and perform, why do we look away from it if it is always there, and how can we expose it without idealizing white expressions of care?

Calibrating the lens

There is nothing original in ending a story with light on the horizon, characters facing or moving towards it, the light flashing as a promise for a brighter future.

According to Toni Morrison, the "light" motif, which happens to be a symbol of whiteness, concludes a great number of American fictions (*Playing* 32–4). Don De-Lillo's most recent novel, *Zero K*, makes no exception. At the end of a rather bleak story, the main character, Jeffrey, stares at the light ahead. This is a promising ending for a man who has looked for answers throughout the entire story, and who has seen the darkness of a dying, war-torn world. Armed conflict may not be resolved in *Zero K*, but there is still hope.

A cult of scientists has discovered a cure for death. For Jeffrey, light flickers stronger than hope, and has something godly about it (274). As it turns out, Jeffrey is on the winning side of history. His family is the first to try the new technology and he has a proven capacity to progress and change. Other than witnessing human and earthly decay, Jeffrey remains safe and protected throughout the story. As a visionary facing the light, he testifies to *Zero K*'s optimistic finale and affirms the power and persistence of hope. Symbolized by light, hope appears bright and unified on *Zero K*'s last page. Here, hope is a spectrum of colors, the entire range of wavelengths combined, shining a bright and heavenly light:

> We were in midtown, with a clear view west, and he was pointing and wailing at the flaring sun, which was balanced with uncanny precision between rows of high-rise buildings. It was a striking thing to see ... the power of it, the great round ruddy mass, and I knew that there was a natural phenomenon, here in Manhattan, once or twice a year, in which the sun's rays align with the local street grid ... The full solar disk, bleeding into the streets, lighting up the towers to either side of us, and I told myself that the boy was not seeing the sky collapse upon us but finding the purest astonishment in the intimate touch of earth and sun. (273–4)

That light can unify and manifest is *Zero K*'s resolution. Light signifies the possibility of and hope for renewal. It "clears" the story's conflict, bursting open the typical for DeLillo claustrophobic narrative space, and revealing the opportunity for a better, renewed world. It is this light that culminates Jeffrey's journey and settles his angst. Throughout the story, while Jeffrey is still evolving, he sees scattered, indistinct figures. The figures vanish like spectres. Their presence is intermittent, as if light has faded, and one cannot see them, neither do they form a sharp image. The figures emerge abruptly in the story, startle Jeffrey and *Zero K*'s readers, and build suspense softly, reticently. Flat and undeveloped narratively, the spectres seem to bear no relevance to the plot. Contrary to the light seen at the end of the novel, they do not align for a stronger meaning.

The spectres are shadows who materialize and perish suddenly, seemingly without influence on the linearity and coherence of the story. Although abstract and vague, I take these spectres to be read as racialized, Black*end bodies directly opposed to the white characters in *Zero K* and to the white light these characters see. In the following, I will discuss how white light, or what I argue to be idealized and ideological whiteness, arises in juxtaposition to the seemingly unimportant spectral Other. In my view, the tension between accentuated and clear white(ness) and the spectral Other is crucial to the story. Focusing on this tension, I direct my study towards a rare consideration in Don DeLillo's oeuvre: DeLillo's complicity in anti-Black representation.

Literary criticism has paid little attention to questions of race and racism in DeLillo's novels (Row 84), treating them like the narrator treats the spectres in *Zero K* – one acknowledges their existence but deems them irrelevant to the larger picture. In contrast, my analysis draws attention to the commonalities of writing and reading white heroes – considering both DeLillo's fiction and him as author/ ity – and how they obscure the concurrence of a racialized Other. In white American literature, Morrison argues, characterization depends on an antagonism between positive whiteness and the obscure, vilified role of the Black (*Playing* 6, 17). Examining the ambiguous presence of Blackness in *Zero K*, I follow Morrison's theorizations and discuss how this antagonism enables DeLillo's narrative construction. A closer look at the novel's reception reveals that the antagonism underpins not only literary texts like *Zero K* but reading practices at large.

Although considered controversial, DeLillo is often celebrated, much like the main hero in master narratives about heroic universality. As Jess Row points out, DeLillo is seen as

> a writer who has escaped race and locality, who never betrays "ethnic consciousness," who has become perfectly national, plural, "essential" – over a body of work that obsessively picks apart the certainties these narratives require. (86)

Zero K's men take center stage under similar lighting. Far away from the spotlight of narrative focus and critical attention, DeLillo's narrativization of Blackness seems minor and sketchy. Appearing as nothing but shadows in *Zero K*, Black bodies are also overshadowed by the light falling elsewhere – on (the) white author/ity. I pause on this contrast and investigate how it blinds the reading eye for brazen and buried instances of anti-Blackness in *Zero K*. My interest lies in the shift of light – the way light moves to create key personages, evades certain spaces and bodies, and transfers from structures staged in the text to practices of reading and modes of understanding. I begin, therefore, by contextualizing DeLillo's image and examining the ties between his reputation and existing interpretations of *Zero K*. My analysis suggests an overlap between the ways lighting has worked in *Zero K*'s narrative construction and DeLillo's canonization. As a technique regulating which figures appear as central and which as marginal, lighting accentuates and conceals nuances and features, and controls the viewer's focus. I am interested in the organization and effects of this technique, and to this end I will analyse some of *Zero K*'s most cherished characteristics and characters and what DeLillo and his critics often relegate to the margins.

Don DeLillo

Don DeLillo has entered the literary canon with the recognition that his works are a highpoint of creative experimentation and a deeply political engagement with the urgencies of U.S. culture and history. It takes just a brief look into the long list of awards and nominations Don DeLillo has received over the years to understand the place he occupies in the contemporary literary world. A 2016 conference organized in Paris and dedicated to the writer might reveal just as much. It ran under the banner "Fiction Rescues History" and aimed to address, among other things, DeLillo's insurgent poetics against official history and its oppressive protocols. To me, the title of the conference and its spirit summarize to a great extent a widespread opinion of DeLillo and his oeuvre. As Charles Baxter notes, the "typical DeLillo tale reads like a diagnosis of a zeitgeist malady we never knew we had [and] DeLillo has achieved greatness that [is] for a contemporary American writer, unsurpassed" ("A Different Kind"). DeLillo's capacity to diagnose is often attributed to visionary and ethical reflections on contemporary crises.

For Duvall, "(w)hat makes DeLillo one of the most important American novelists since 1970 is his fiction's repeated invitation to think historically" (2). Back in 1991, Frank Lentricchia connected DeLillo's historical awareness to cultural criticism and social liberation and named him one of the writers "who invent in order to intervene" (2). Similarly, Daniel Aaron describes DeLillo as a "sociologist of crisis" (70). This feeling has underlain many of the appraisals the author received over the years. As Duvall underscores, DeLillo was awarded the Jerusalem Prize in 1999 for "his fiction's moral focus [and for] wrestling a bit of freedom from necessity by so thoroughly diagnosing what constrains us" (3). Thus, DeLillo is often recognized as a writer whose creative intentions and achievements rescue, at least conceptually, American history.

A recurring description, a sentence Robert Towers wrote in *The New York Review of Books* in 1988, positions DeLillo as "the chief shaman of the paranoid school of American fiction." The idea of the centrality and healing capacity of his work, as well as its preoccupation with social concerns, has braced a staggering amount of literary criticism and mainstream recognition. The idea also redoubles in reviews of *Zero K* – DeLillo's most recent novel. For Joshua Ferris, the book reaffirms DeLillo's "individual genius" and his place as "the seeker, the prophet, the mystic, the guide" ("Joshua Ferris Reviews"). Ferris also deems *Zero K* one of the finest novels of the 21st century, and with this joins critics and writers like Michiko Kakutani, Charles Finch, Sam Lypsite, and Kevin Nance.

Like other DeLillo novels, *Zero K* is praised for thematizing "eternal human concerns" (Kakutani) and for being historically conscious. Marketed as a "deeply moral book" (Levin), *Zero K* is seen as self-reflexive and inspirational. James Kidd notes, for instance, that it is "a visionary novel of ideas that remembers even visionary novels of ideas are read by living, breathing humans" ("Don DeLillo's Latest"). Mirroring this rhetoric, *Zero K* thematizes such idealization. As Graley Herren points out, DeLillo has frequently ironized (the) male author/ity by embedding himself in his fiction (*The Self-Reflexive Art*). To celebrating *Zero K* as self- reflexive leads also the reception of DeLillo's novels as "historiographic metafiction": often popular works which parody dominant discourse, rely heavily on historical events and personages, and subvert official versions of History (Hutcheon 114). The following analysis shows, however, that rather than ironize and dismantle dominant representations, novels like *Zero K* often reinforce them, uphold the status of (the) popular author/ity, and highlight images always already fixed "on the retina" (Sharpe "Response").

Zero K

Zero K tells the story of the American billionaire Ross Lockhart and his son Jeffrey, who meet to accompany Ross's second wife, Artis, to her artificially induced death. The two men are positioned as the all-American heroes: Ross is powerful and rich, while Jeffrey demonstrates his imaginative depth and values through ethically-charged inner monologues and personal progress. As in other DeLillo novels, the characters inhabit a claustrophobic, alienating world full of destruction and terror. In *Zero K*, this world is torn by war and the human body torn by death. Artis is terminally ill and about to be frozen for a distant future wherein she will be revived. Artis is kept at the Convergence where an elite society of white men uses new technologies to achieve human transcendence and renewal. The overarching theme of the novel is the future return to a better life – a reinvented world and body, and a new consciousness to take control of both. The protagonists explain early on:

> "Yes. The time will come when there are ways to counteract the circumstances that led to the end. Mind and body are restored, returned to life."
>
> "This is not a new idea. Am I right?"
>
> "This is not a new idea. It is an idea," he said, "that is now approaching full realization." (8)

The novel's main interest is renewal and, at first glance, all characters seem to be striving towards it (71). A close reading reveals, however, that not everyone in the story gets to rid themselves of mortality and thus be reborn – some are more capable of dying, others are the cause of death, and a chosen few are destined to rewrite the future and experience rebirth/renewal. The roles and repertoires are cast according to characters' race and gender: white men progress and move forward, white women escort them during this journey, and Black bodies are both nowhere and a threatening unknown. This detail has been widely ignored in literary criticism. On the one hand, the misrecognition can be linked to postmodern techniques like parody and irony. As ironical, the text can be said to expose contemporary structures in their exaggeration (see Hutcheon, Shugart). Thus, *Zero K*'s portrayal of female bodies as the perfect object may be interpreted as a comment on the longstanding traditions of women's objectification, while the textual exclusion of Blackness may function as a comment on exclusionary politics. Another explanation for literary criticism's lack of interest in *Zero K*'s racialized and gendered tensions is the fact that the story continuously conceals them through other urgencies and plots. This, in turn, disqualifies irony or at least puts into question its overtness.

I mention irony's overtness here because, at its heart, irony "has a corrective function," it is

> like a gyroscope that keeps life on an even kneel or straight course, restoring the balance when life is being taken too seriously or, as some tragedies show, not seriously enough, stabilizing the unstable but also destabilizing the excessively stable. (Muecke 4)

In order to expose imbalance and correct it, irony requires to be overt – if not in its expressive form at least in its effects. As we will see, *Zero K*'s gender and racialized tensions fail to undo themselves even if we read them as ironic. These tensions carry (out) dominant representations and strengthen rather than destabilize them. One way to put the question here is whether irony is sufficient or successful enough. Yet, this perspective limits the analysis to DeLillo's uses of language. Are the tensions missed because they are ironic, or because they are concealed, i.e. not ironic enough? In other words, are critics misrecognizing them because of the work of a particular text? Or, is the question of misrecognition connected to discourse, to the much larger issue of authority and its navigations of meaning?

I argue that both ironizing and concealing these tensions fails to account for the general misrecognition of *Zero K*'s racism. I follow Black critique in suggesting that, the misrecognition arises from the fact that narratives of renewal have long depended on Black death and destruction (Wilderson, *Afropessimism* 225). As Calvin Warren asserts, "black suffering and death [are] the premiere vehicles of political perfection and social maturation" ("Black Nihilism" 7). Conceptualized as the American Dream, perfection and maturation fuel movements towards the renewed, better state (see Bercovitch, Cullen). Yet, "the [American] Dream rests on [Black] backs, the bedding made from [Black] bodies" (Coates, *Between* 11). From this perspective, silence over DeLillo's racialized representations sounds logical and in consensus with dominant histories of interpretation. They are so normal, so commonplace for white visualizations of renewal that the white gaze hardly needs to pause and register them. Instead, white readership brushes past. Ironical or not, the effect of dominant representation is its normalization, its continuous re- entering into discourse as an insidious, invisible regularity (see Hall, Stam, Mirzoeff).

There is another point I want to underscore here. When Frank Wilderson writes that it is "the incoherence of Black Death [that] generates the coherence of White life" ("The Prison Slave" 232) he points to the relationship between white quests towards renewal – the claim on meaningful existence – and the negation of Black being. To celebrate *Zero K* as a visionary and ethical masterpiece without critically reflecting its racist cultural work is not paradoxical in a world where Black exclusion always already goes hand in hand with white dreams for better futures (see *Futurity*). In fact, racialized tensions in the novel cause no damage to its narrative coherence (because they reiterate dominant discourse in intelligible ways) and to its critical reception (because visions of a better world, in their white configurations, are unproblematically anti-Black). From this vantage point, overlooking *Zero K*'s racism owes to the ease of reading with and past anti-Blackness. The following analysis does not so much unearth the latter in *Zero K*, as it demonstrates its brazen and simultaneously "buried" presence(s), the failure of literary forms to deconstruct these, as well as the relations between white ideology and portrayals of renewal.

Zero K's men

The trope of evolving white men headed towards renewal drives the plot. Pertinent to another project is the investigation of *Zero K*'s alignment of all male characters with holy historical figures, and divinity. I want to note briefly, however, that the men's depiction in *Zero K* unambiguously frames them as the proverbial forefathers.[1] Jeffrey's father, Ross, signifies, the godly and almighty ancestor, while Jeffrey,

---

[1] As Alan Parker summarizes, "DeLillo has given us a new way of thinking about the old guy in the hooded cape" ("Mortal Panic")

in resemblance to Jesus (and arguably, Thomas Jefferson), witnesses the destruction on earth and rises above those who sin. The father and son join two brothers, the Stenmarks,[2] who invent and safeguard the future, and with this reinforce the image of white men as creators, saviours and visionaries. A close reading lucidly explicates this three-fold assembly. On the one hand, the triad characterization is strengthened by staging the father and the son alongside another man in story: the Monk. The three of them figure as the holy trinity and further establish the context for renewal and rebirth. Not accidentally, for instance, does Ross leave Jeffrey the moment he writes "over and over: *sine cosine tangent*" (14, emphasis in original). Signalling the relationship between the sides of a triangle, the phrase functions as a leitmotif throughout the text and underscores the blessed men's connection to each other (67, 123–124).

Although Ross disappears and leaves his son to grow up and witness the destruction of the world alone, he awaits Jeffrey at the Convergence, the hopeful space where the Stenmarks attempt human rebirth (7). *Zero K*'s numerous religious references and associations with important figures further brace white men's image as holy and eternally progressive. Three categories strengthen this characterization: men's capacity to 1) see/envision/discover, 2) evolve/grow/progress, and 3) create/survive/originate. Although these capacities are directly linked to white ideology in the story, they remain conveniently misinterpreted in literary analyses. Instead of ironizing white male authority, everything male characters can do is used as evidence for DeLillo's own capacity to surpass the present.

To see why critics focus on DeLillo's inspirational tale rather than on the story's recourse to ideology, we need to look closer at Jeffrey's use of the capacities and their relation to renewal. One the one hand, renewal is motivated by Jeffrey's emotional adolescence. To Ross, he admits: "I'm doing my best to recognize that you're my father. I'm not ready to be your son" (113). Jeffrey is described as unprepared, wondering, and eager to learn. The shifts between jobs and women, and between the fullness and abridgement of his own name (18), mark the fluidity and incompleteness of Jeffrey's identity. Contrasted to Ross's stability and dominance, Jeffrey appears before us boy-like, confused and curious (57, 115). He mumbles, shuffles, and questions incessantly – all childish things to do – and, yet, Jeffrey is "eager to be educated"

(93) and encouraged to go beyond his limitations. Jeffrey's openness to change and growth are not obscured in the story. Instead, they appear matter-of-factly:

---

2 The name is likely a reference to the Stenmark twins, two renowned Australian-born models who built a modelling career around their looks and identical genes (see https://stenmark.life)

I saw myself in the limp, in the way I refined and nurtured it. But I killed the limp whenever my father showed up to take me to the Museum of Natural History. This was the estranged husband's native terrain and there we were, fathers and sons, wandering among the dinosaurs and bones of human predecessors. (107)

Jeffrey literally learns to walk straight – he evolves, in the space and continuance of natural history, following his father(s). Importantly, the depiction of the museum as "native terrain" stages men's evolution as inherent and natural. While there is a lot to unpack here,[3] I want to point out that indications of ever-evolving white men, (social) Darwinism and racial superiority, are already anticipated a sentence before Jeffrey's remarks on the limp. He says: "*Bessarabia, penetralia, pellucid, falafel.* I saw myself in these words. I saw myself in the limp" (107, emphasis in original). On the one hand, the words trigger cognitive maps leading to the Ottoman empire (and its people) and summons stereotypes[4] of the Oriental other (Said 41). On the other, they signify Jeffrey's limp – the inability to walk (straight). It is this implied backwardness that Jeffrey abandons in the museum. This is not a new, or exclusively DeLillo's, story. In a letter to William Graham, Charles Darwin writes that the "more civilized so-called Caucasian races have beaten the Turkish hollow in the struggle of existence" (316). Clear to me seems the echo of this racist estimation in Jeffrey's faculty to walk within natural history.

Positioned in constant movement – we find Jeffrey in planes, buses, and taxis – Jeffrey is also described as an explorer. Moreover, he persistently presents himself as a tourist (62, 185) and as someone who can help others by going somewhere (36, 226). For instance, Jeffrey arrives at the Convergence to console Ross because Artis is dying (3), he is escorted from place to place to understand the purpose of the Convergence (78, 90), and goes to his desperate girlfriends when they disappear and require help (226). In fact, Jeffrey travels convinced that he is needed, that he needs to do something. His journey is motivated by an urgency to be educated (93) and the desire to truly see (140, 153, 204, 258). Thus, curiosity seems inherent to Jeffrey's character and in no way circumstantial. For instance, Jeffrey is obsessed with making up definitions (55, 59, 103, 109, 151). "This is what I did in any new environment," he explains, "I tried to interject meaning, make the place coherent"

---

3 Note that in this quote women are doubly absent – firstly, they are not in the museum, and secondly, their absence is indexed by a textual refusal to name them. In other words, women are absent in the story and in the text telling this story (this is important for arguments I put forward later).

4 "Bessarabia" and "falafel" mark the specific location, whereas "penetralia" and "pellucid" imply this location's secrecy and simplicity. As Said argues, the othered space is imagined as dangerous and inferior (41). Conflating the othered space with the "limp," Jeffrey frames it as backward and disabling.

(10). Combined with the capacity to define and render things meaningful (or not), Jeffrey's movements highlight his resemblance to explorers. He frequently names and renames characters and places – an authoritative gesture practised by Europeans especially during the Age of Discovery. A parallel between Jeffrey and Captain Cook further solidifies this image:

> She gave me a wristwatch and on my way home from school I kept checking the minute hand, regarding it as a geographical marker, a sort of circumnavigation device indicating certain places I might be approaching somewhere in the northern or southern hemisphere depending on where the minute hand was when I started walking, possibly Cape Town to Tierra del Fuego to Easter Island and then maybe to Tonga. I wasn't sure whether Tonga was on the semicircular route but the name of the place qualified it for inclusion, along with the name Captain Cook, who sighted Tonga or visited Tonga or sailed to Britain with a Tongan on board. (107–108)

As the explorer, Jeffrey follows a compass, "checking the minute hand," and travels around his miniature-world. Note that the wristwatch locks certain geographical places in time. By imagining Cape Town, Tierra del Fuego, Easter Island and Tonga inside the watch, Jeffrey frames them as his property. Here the narrative echoes his/stories of Western explorers who discover a world that is, in their eyes, static and outside progressive time. Again, this is not DeLillo's invention, but a reiteration of a longstanding mythology (Adas 117). Its repetition frames Jeffrey as the explorer, on the one hand, and multiplies the circulation of narratives of western conquest, on the other. As Mary Louise Pratt argues, these narratives depend heavily on portraying white explorers in opposition to an "inferior" population (183). To a great extent, this inferiority is described through representations of othered bodies as outside culture and progressive time.

Another slippage the passage "works" for is the conflation between Tonga-the-land and Tongan-the-person. The collapsing of sighting/visiting Tonga and bringing a Tongan on board is marked by the interchangeability of these events. Put otherwise, the uncertainty and unimportance of whether Cook went to the land or returned with the person contributes to the abstraction of Tongan personhood, or what Aimé Césaire calls "thingification" of the colonized (42). In DeLillo's representation, Tonga and its population become a homogeneous mass of "Tongan-ness," where space and people are reduced to a unified idea, and thus rendered indistinctive. It is important to note that in this implication, the passage emerges as coherent because racism and reductionism are already at play. As Christina Sharpe points out, contemporaneity is always already anti-Black and the reality of anti-Black practices seen as normative (In the Wake 104). Similarly, Martinot and Sexton clarify that "racism is a mundane affair" (173). They write further:

> [w]hite supremacy is nothing more than what we perceive of it; there is nothing beyond it to give it legitimacy, nothing beneath it nor outside of it to give it justification. The structure of its banality is the surface on which it operates. Whatever mythic content it pretends to claim is a priori empty. Its secret is that it has no depth. There is no dark corner that, once brought to the light of reason, will unravel its system … its truth lies in the rituals that sustain its circuitous contentless logic; it is, in fact, nothing but its very practices. (175)

As normative, "mundane affair," the racist implications of this passage figure as a shortcut to a dominant episteme. There is an available understanding of white superiority which rationalizes Jeffrey's character. *Zero K*'s ideological markers only direct us – full circle – back to it. Rather than ironize who and how they get to move in the text, however, Jeffrey's character enhances the image of white authority. Thus, it is Jeffrey's character which highlights the narrative's geographic specificity and allows for interpretations of the story in terms of progressive transnationalism. For instance, Jeffrey's movement marks distance and the characters' dis/connection to/from far-off places – North America, Tibet, China, India, Switzerland, Germany, England, Kazakhstan, Iran, Ukraine, Russia. Jeffrey's movement thus indexes the (diegetic) world's wide terrain. It is important to note here that geographical specificity and expansiveness are often read as DeLillo's reflection on the border concept. M.C. Armstrong writes, for instance, that DeLillo's

> celebration of the self-interested spirit suggests that, rather than seeing history as a dialectic of classes or collectives or discourses or tribes, it might be framed as "single lives in momentary touch." ("Back")

Similarly, frequent references to (missing) nationalities and passports imply *Zero K*'s global landscape and movement (19, 27, 63, 113, 169). Like Armstrong, Jeffrey asserts that cultural difference falls off:

> Here I was, in a sealed compartment, inventing names, noting accents, improvising histories and nationalities. These were shallow responses to an environment that required abandonment of such distinctions. (72)

In fact, the story welcomes its interpretation in the context of border regimes and transnationalism. The characters frequently lose their names and languages, Jeffrey often comments on illegality and on other characters' countries and foreignness (19, 169, 268). Thus, when Jeffrey says that they "are completely outside the narrative of what we refer to as history [and that there] are no horizons [there]" (237), a portrait of a global landscape unravels. This globality triggers praises of *Zero K* as a depiction of a postracial and a post-postcolonial world, or at least at their narrative anticipation. Joshua Ferris remarks:

We are in a vision of the future, a postracial, post-postcolonial world where Westerners like Ross and Jeff are but one contingent of a technocratic cult with a single aim: to rid the world of that absolute, all-defining force, that ultimate despotic colonizer, death ... But the seduction is every bit as illusory ... After all, we are as far from a postracial, post-postcolonial world as we are near to arriving at whatever technique or technology is necessary for eternal life. ("Joshua Ferris Reviews")

Ferris' statement is problematic for two reasons: on the one hand, the statement is problematic, because it frames colonization as a plight for the living, i. e. exclusively white, characters in the story. Thus, Ferris further suspends Black being – the bodies who suffered from colonization are removed from interpretations of *Zero K*, the white characters are dissociated from historical responsibility and from their position as oppressors, and the structure of colonial violence is used as a metaphor which grants narrative tension rather than reflect historical actuality. The statement also suggests that *Zero K* exists on the same imaginative plane and in the same progressive linearity which ties historical awareness – DeLillo's depiction of the imperfect world – to the prospect of social transformation. In other words, the postraciality and post-postcoloniality of the future is somewhere out there, just not "here" yet. However indirectly, the contention entails that narratives like *Zero K* lead to, or at least do not sabotage, the possibility of a postracial and post-postcolonial future. On the contrary, they allude to its existence or even lead to it by way of envisioning or pointing towards it.

The unchecked normativity of who survives, evolves, and reinvents themselves in *Zero K* – exclusively white characters as this book underscores – renders Ferris' observation paradoxical. Both the future world in the narrative, and the text itself, bar living Black bodies. Rather than eliminate racism and racial violence – a certain prerequisite for a postracial and post-postcolonial world – the story eliminates Black being.

Let us assume that DeLillo ironizes white authority in *Zero K* which is also the assumption underlaying Ferris' statement. For the reasons mentioned above, the problem that irony aims to expose or destabilize is raised again: by the practice of misrecognizing/ignoring anti-Blackness and by highlighting/idealizing white author/ity. Reflecting dominant discourse, DeLillo's representations are guided back to it. In other words, even if the representation is ironical, it cannot escape white meanings because they seep in and out of what we see, always already working for the reinvigoration of whiteness.

Therefore, the issue is not restricted to racist images but rather extends to racist imagination. As Hortense Spillers says, racist representations are "so loaded with mythical prepossession that there is no easy way for the agents buried beneath them to come clean" (65). *It takes a special kind of ignorance to praise white author/ity for washing off racist images by re- using them, and consider* *this cleanse a confirmation for white interest in postraciality. The idea that white author/ity is working towards the greater good, towards some universal renewal, is so strong that it brushes past **[jumpcuts]** the repetition of transgression even when it aims to focus on this transgression.*

Objects in the distance

For a story imbued with geographical and cultural specifics, *Zero K* forgets to include Black characters and only fleetingly refers to Black spaces. The appearance of South Africa, the Bronx, and Egypt in the texts rather contrasts, then partakes in the story's topography. These geographical markers do not function as the other territories in the story. Instead, they emerge abruptly and carry no significance. The Mastabas, which are the reason for citing Egypt, are only the model for a building, which is not in Egypt itself (41). Cape Town is referred to in a prolonged sentence describing Jeffrey's movement as an explorer (107). South Africa surfaces in another complex sentence, this time indicating the movement of a guy's gaze: "he droned his way through a global roundup that ranged from Hungary to South Africa, the forint to the rand" (165). In a similar structure appears the Bronx:

> The subway is the man's total environment, or nearly so, all the way out to Rockaway and up into the Bronx, and he carries with him a claim on our sympathies, even a certain authority that we regard with wary respect, aside from the fact that we would like him to disappear. (58)

In all cases, the reference to location aims to highlight distance. Moreover, the characters are never in, but rather relate their movement to, Black spaces. They thus appear as nothing other but a marker – uninhabited, irrelevant. This emptiness redoubles in the story. As mentioned before, Black characters are not named, and this, together with the voiding of full/lived Black space, works for a revised (and racialized) globality. This narrative manoeuvre also strengthens what Katherine McKittrick exposes as white practices of rendering Blackness "ungeographic." As McKittrick writes, "the legacy of racial dispossession underwrites how we have come to know space and place" (4), and white conquest, land ownership, and colonization are directly linked to the present economy of imagination that positions Black bodies outside the normative world and its specificity, variety, and full-ness. *Zero K* reproduces this Hegelian erasure – it situates Africa outside progressive history, abstracts Blackness, and fixes Black bodies as quintessentially dead. This is further suggested by the way the text excludes Black bodies but consistently forestalls Blackness's presence. In its configurations as a body-less and gender- less entity, Blackness haunts Jeffrey and gives the story acceleration to unravel.

A straightforward way in which Blackness enters the text is the Black object. For example, there are numerous references to Indigenous and Black styles and things: Jeffrey has an "aboriginal shaved head" (15). A nameless woman wears "a black Navajo hat" (214), another an "Arab headdress" (179), and yet another one has "hair bunched afro-style" (236). Jeffrey drinks Portuguese Madeira (205); a man wears a "safari jacket" (230). Blackness is introduced through the things white characters use and are in possession of. In fact, the characters" whiteness intensifies the feeling that the Black body is not there and only the Black object remains as a signifier. This contrast is also implied by references to deserts, jungles, and Mastabas, and by DeLillo's persistent refusal to name Black people in these locations. Jeffrey copies this gesture by resisting "an impulse to name [a woman he meets] like a color" (23). Although, he encounters several people with a "dark face" (19), they are never really Black.

Another example for absent Blackness is the human skull Jeffrey sees at the Convergence. The skull unambiguously marks white settler colonialists' practice of exhibiting African bones, and the racial stereotypes that came with, and inspired, such representations. In *Zero K*, the skull is over-sized which is a direct replication of widely spread scientific racism (and phrenology more specifically):

> It was about five times the size of an ordinary human skull and it wore a headpiece, which I hadn't fully registered earlier. This was an imposing skullcap in the shape of many tiny birds, set flat to the skull, a golden flock, wingtips connected ... It looked real, the cranium of a giant, blunt in its deathliness, disconcerting in its craftwork, its silvery grin, a folk art too sardonic to be affecting. I imagined the room empty of people and furniture, rock-walled, stone-cold, and maybe the skull seemed right at home. (68)

The point I want to push here is that objects in the story imply Black presence but underline its bodily absence. Thus, Blackness is summated as abstract. It is implied outside the precincts of country, context, and characters, and emerges as unknown and larger than body thingness. As Ferris underscores, what threatens *Zero K*'s characters in the story is an abstract enemy – death. In many ways, the story allows for taking an imaginative shortcut: following dominant discourse, the story's abstracted Blackness sits near enough to the abstract threat that kills white women and that white men set out to defeat.

It is important to note here that Ferris' imaginative manoeuvres are neither far-fetched, nor contextually isolated. On the contrary, DeLillo's image as a historically conscious writer motivate them, as do certain images in *Zero K*. For instance, by naming each part of the story after zones of contact/collision,[5] DeLillo not only places his characters in a world that spans geographically but demonstrates an engagement with contemporary crises. Yet, neither the reference to collapsing/colliding worlds, nor Jeffrey's characterization ironize white men's quest and authority. Sympathetic to Jeffrey's desires to grow and become better, the story reaches its resolution: Jeffrey becomes "an ethics officer" and finally realizes the potential to supervise and oversee (221). What he sees is no longer destruction and chaos. On the contrary, at the end of the novel Jeffrey stares at the bright, heavenly light. This light opens the horizon and marks the positive rise of possibilities. It is this moment that bridges Jeffrey's movement forward and readerly hope. At this point, Jeffrey's authority remains intact because he survives as a moral hero not only in the story but in its subsequent receptions as well. Thus, Rachele Dini uses Jeffrey's progressivity and moral impetus to conclude that in *Zero K* life extension is a form of bodily narration:

> The ultimate marriage of art and matter, it amounts to "telling" ourselves to continue living, to continue existing beyond what we thought was the story's end, to indeed apply the rules of the novel to create the future itself – not only its representation. (2)

Like Ferris, Dini links Jeffrey's progressivity and possible futures to DeLillo's capacity to narrate and bring us, the readers, forward. Yet, to prescribe authority to the characters who have voice and participate in world-making (white men like Jeffrey) follows a dominant ideology: active and mobile men are not ironized or deconstructed but rendered inspirational. This might appear as an odd conclusion – after all, DeLillo is famous for his bleak landscapes and characterizations, and his novels might be described as depressing. Similarly, *Zero K*'s finale might be optimistic and epiphanic, but hardly compensates for the nagging feeling of isolation the story – and DeLillo's oeuvre in general – work for. Yet, Dini is not alone in drawing inspiration and hope from DeLillo's fiction. On the contrary, the novel's reception shows that we may read past *Zero K*'s concluding sentence and into fictions around DeLillo himself. As a writer who narrates, or diagnoses, contemporary crises, he seems to inspire white people's own potential in re-imagining future existence; his voice and author/ity to tell are not ironized but deemed positive, constructive, and powerful. Maddie Crum writes

> [*Zero K*] and it's life-affirming conclusion, could be read as a triumph of honest language, the kind of human expression that comes out unfiltered when we're spurred on by awe, by ur-

---

5 Part One is named after the meteor which descended over the Ural Mountains in February 2013; Part Two is named after the city were Russia and Ukraine clashed in 2014 (2,163).

gency, by the promise of eventual death ... As ever, DeLillo explores the depths of an edgy, timely topic, completely resisting cliché, and emerges with something both fresh and universal. ("The Bottom Line")

It should be noted here that *Zero K* thematizes contemporary crises deemed such by leftist circles (e.g., environmental collapse, patriarchal oppression). Can the novel be praised for highlighting leftist ethics, however? To me, Ferris' reference to postraciality and Crum's reference to universalism are connected to such engagement with topical issues. They are also connected, however, to DeLillo's mainstream reception – i.e., DeLillo's image influences interpretations of *Zero K*'s narrative space. In fact, narrative space and how it appears before us – as something other than what it is – is regulated prior to interpretation. By centring questions of bodily and earthly decay – a focus that presupposes gender and environmental concerns – *Zero K* pushes back or "buries" the moments in the text where renewal is described as an exclusively white prerogative. What is more, *Zero K*'s continuous references to post-identity and post-ideology conceal the fact that the story, at its base, depends on racial antagonisms.

As the next sections shows, however, racial antagonisms are central to the story and work to highlight white author/ity as the owner of renewal. Therefore, *Zero K* departs from leftist interventions,[6] and fails to ironize the key images it employs in the self-reflexive run on/in History. This becomes clear if we consider the function of some of the most obvious ironic representations in the novel – female characters and gender issues in *Zero K*. Described through blunt stereotypes, white women take central stage (not textually, but in men's thoughts in the story). The tension between men and women is so accentuated, in fact, that gender trouble seems crucial to the plot. Yet, women's presence, and gender themes respectively, conceal more insidious characterizations: like the association of Blackness with danger. After the opposition between men and women in *Zero K* serves the plot, it is abandoned – women disappear, and only white author/ity and vilified Blackness remain.

---

6 For example, Octavia Butler's *Xenogenesis Series* which thematize many of *Zero K*'s concerns in a drastically different way than DeLillo's novel. An interesting comparative study can be made between Butler's *Dawn* and *Zero K* – I leave this analysis for future engagements with the racialization of narrative and language

Bicycle

To understand how white author/ity can appear as inspirational rather than problematic, i.e. ironized, it is necessary to examine white characters' overall function in the narrative. Nowhere is this functionality better articulated than in the relationship between white men and women in the story. Here is what Jeffrey observes:

> Now, somewhere else, another town, another time of day, a young woman on a bicycle pedaling past, foreground, oddly comic motion, quick and jittery, one end of the screen to the other, with a mile-wide storm, a vortex, still far off, crawling up out of the seam of earth and sky, and then cut to an obese man lurching down basement steps, ultrareal, families huddled in garages, faces in the dark, and the girl on the bike again, pedaling the other way now, carefree, without urgency, a scene in an old silent movie, she is Buster Keaton in nitwit innocence, and then a reddish flash of light and the thing was right here, touching down massively, sucking up half a house, pure power, truck and barn squarely in the path ... Total wasteland now, a sheared landscape, the image persisting, the silence as well. I stood in place for some minutes waiting, houses gone, girl on bike gone, nothing, finished, done. (36–37)

In the passage above, the girl is foreign, far away, nameless, small, innocent, and soundless. The comparison to the American actor Buster Keaton relates her to his silent movies but also to his career as a stunt performer, comedian, and screenwriter. Thus, she emerges before us in her "oddly comic motion" and invokes silence, role(playing), and fiction. These allusions redouble her description – the movement on a bicycle summons our memories of Keaton. He appears together with her, amplifying the image. Her foreignness is redoubled, too. Not only is she far away by appearing on a screen (Jeffrey watches her from a distance), but she is in "another town, another time of day" and she is distant. Like the far-off vortex "crawling up out of the seam of earth and sky," she is "quick and jittery," a double of its remoteness and size. Her smallness repeats also in the opposition to an obese man who happens to be "ultrareal." Again, his realness reiterates in the opposition to the girl's un-realness: she is merely a resemblance of Keaton in a screen within a screen, "a scene in an old silent movie."

The girl emerges suddenly and is suspended in the hallway. She disappears suddenly and is suspended in the image in turn, swallowed by ruins, "half a house ... in the path," her motion stopped midsentence. Like other female characters, she becomes recognizable in her resemblance to an actress. Conceivably, Sybil Seely, one of Keaton's famous colleagues, is implied here. Seely and Keaton partnered frequently, and often appeared together, as they do in *Convict 13*, *The Scarecrow*, *The Boat*, and *The Frozen North*. Moreover, Seely herself replicates the girl's characterizations of comic-ness, innocence, and she emerges on screen as well.

Thus, in the bicycle passage the girl aligns once with Keaton, and once with the actress. A third time she aligns with both in another, widely circulated picture – it is Keaton and Seely together, riding a bike.[7] In this image, Keaton stares at Seely who is sitting and, distracted, looks somewhere else; Keaton rides the bike for both of them and carries Seely forward.

The point I want to make here is that the narrative buries an image within an image. It thus connects different representations – Keaton's movies, for example. It amplifies them and makes them easier to imagine. Thus, the narrative re-enters imagination because it is already there. There is an underlying story that is solidi-fied through the discussed repetitions. I argue that, self-reflexively or not, DeLillo extends white mythology here and re-creates and re-buries female characters for a reason. Or, the girl appears (to disappear) not so much to underline female inferi-ority and thus diagnose gender troubles. Rather, she, like the other female charac-ters in the text, delivers the contrast between 1) righteous whiteness (reinforcing the tropes of the white saviour and the damsel-in-distress) and 2) abstract evil by indexing the existence of a threat against her. Once again, this idea does not owe to DeLillo's originality. As Bogle points out, whiteness has continually employed the tropes of the innocent/endangered white woman, of Blackness as intrinsic/uncon-trollable evil, and of white men as saviours (10–15). Replicating this representa-tion, *Zero K* continuously depicts white female fragility, desired sexuality, and pu-rity.

Like most female characters in *Zero K*, the girl vanishes – but only after she has indicated her worthiness to be loved. Threatened by (natural) disaster, women are the purpose, the fuel to men's quest. Jeffrey's progress happens against a background of appearing and disappearing girlfriends, Ross tries to save Jeffrey's stepmother, and the Stenmark brothers work towards her rebirth. Continuously de-scribed as escorts, women in the story satisfy men and bring them to their desired destination. Yet, women are never where needed (19, 36, 210, 215, 226) and always seem to be in the middle of something (9, 258). Thus, Jeffrey's stepmother, Artis, "doesn't know how to get out of words into being someone" (157) and eventually turns into a "baby boy" (48). Consisting only of words, she appears as pure fiction (157–62).

Like the girl on the bicycle, women lack sight but are suspended in space to be seen. They are empty and "hollow-bodied," merely objects of admiration (94, 72, 157). One of Jeffrey's girlfriends is so paper-like, so "tall and thin she [is] foldable"

---

7 Similar image is widely circulated online; I have not managed to trace the copyright owner and obtain permission to reuse it but the image can be easily found under the search entries: "Keaton," "Seely," and "riding a bike"

(56). Another one is smeared into the wall as "an imprint, a body mark" (78). Artis, the literal personification of art, reads as foldable, too, as do the other characters positioned artificially in chairs and beds. Furthermore, women appear abruptly out of the background, only to disappear into the background again – with no lineage, merely "Gesso on linen" (268). In the rare occasions when they surface in the text, women are described as deficient. They are nameless, blind, incapable of speaking, always sitting, dying, and barren. At the same time, they are desired. Here immediate focus falls on Artis, the "idealized human" (258), who is of equal importance to Ross and the Convergence, and is most useful in her sickness.

The worship of the female body redoubles in all portrayals. A good example is Madeleine, Jeffrey's dead mother. Framed by two of the central motifs in *Zero K*, misspelling and religion, the character functions as both an escort and a worshipped persona. Thus, her name activates associations with Lady Magdalene and summons popular imaginations of the woman as a repentant sinner, a prostitute, and a holy figure. For instance, Madeleine is both the obliging, "ever-accommodating, self-sacrificing, loving and supporting" woman in the story, and the one who betrays Ross by stabbing him with a knife (100). Although Madeleine has fallen out of grace and lost her name (32), Jeffrey idolizes her and restates her usefulness (15, 248). In this sense, Madeleine mirrors Artis and the idea that a woman can be useful in her death, perfect in her absence:

> Her body seemed lit from within. She stood erect, on her toes, shaved head tilted upward, eyes closed, breasts firm. It was an idealized human, encased, but it was also Artis. Her arms were at her sides, fingers cusped at thighs, legs parted slightly. (258)

Women's deficiency and fragility are their most visible features. Particularly, women lack voice – Artis's language falls off (157–62). Emma does not know how to speak to her adopted son (171–76), another woman has her mind "empty of words" (72), and others "speak" but in "choppy syllable-like units" (230), in vanishing whispers (16, 18, 105, 244). Women also lack names. Jeffrey says of one of them that she "blended better, nameless, with the room" (79), and dismisses another: "[n]ever mind giving her a name" (238). He sometimes names women only to withdraw, misspell, or forget these names later. Although sexualized and marked by eroticized nakedness, women seem robotic, artificial, and never really themselves (157, 48). Each female role is failed by the women who attempt to fill it. The repetition of these images and characters solidifies womanhood as a category and inscribes it in scarcity and sickness.

The explicit derogation with which women are portrayed and which could be misread as irony here has an important narrative function. As it becomes clear in the bicycle passage, female characters are notable in their disappearance. Their sickness and desirability stage them as endangered and in need of protection by white men. Indeed, Jeffrey and the other male characters watch out for women in the story. Yet, gendered characterizations in the text do not sabotage patriarchal forms and formalities. Rather, they lend tension required for narrative coherence. *Zero K*'s women perish but not before indexing their innocence and worthiness to be loved. Thus, their objectification works to redouble the fear that something sinister is out there to harm them (Artis is dying), and to highlight the author/ity of men, the saviours (the Stenmark brothers help Artis reawaken). "Either way she dies," says Jeffrey thinking of his mother, but this conclusion "happens somewhere else," too, and envelopes all female personages in the story (50).

The scarcity of female characters in *Zero K* might suggest that DeLillo could have omitted them entirely. Yet, their (dis)appearances serve the plot. *Zero K*'s women remind readers that there is something to be protected, and that male author/ity is necessary. White women summon the image of the white male hero and the image of abstract danger when their bodies fail, lose abilities, and turn into objects. Objectification serves white author/ity independent of whether it is ironical, i.e. whether was aimed as undoing, or not.

Mannequins

The function of objectification becomes clear in Jeffrey's first encounter with one of the mannequins. He relates this encounter:

> as a figure standing there, arms, legs, head, torso, a thing fixed in place. I saw that it was a mannequin, naked, hairless, without facial features, and it was reddish brown, maybe russet or simply rust. There were breasts, it had breasts, and I stopped to study the figure, a molded plastic version of the human body, a jointed model of a woman. I imagined placing a hand on a breast. This seemed required, particularly if you are me. The head was a near oval, arms positioned in a manner that I tried to decipher – self defense, withdrawal, with one foot set to the rear. The figure was rooted to the floor, not enclosed in protective glass. A hand on a breast, a hand sliding up a thigh. It's something I would have done once upon a time. Here and now, the cameras in place, the monitors, an alarm mechanism on the body itself – I was sure of this. I stood back and looked. The stillness of the figure, the empty face, the empty hallway, the figure at night, a dummy, in fear, drawing away. I moved farther back and kept on looking. (24)

While they will become clearly more threatening later in the text (74, 132), the tension between Jeffrey and the mannequins is already introduced here, as are the first indications for their racialization. Before I unpack this, it is important to mention that these encounters occur exclusively in the liminality of corridors. On the one hand, the corridors contrast with the security and peace of the compound.

Lacking windows, Jeffrey's room does not seem claustrophobic but grants him consolation: in it, Jeffrey is protected from the collapsing outside world (20, 43). Furthermore, Jeffrey sees the building as a clean and cleansing hospice (9, 30, 42, 91). In direct opposition stands the passageway where screens with images of death and destruction startle Jeffrey. The juxtaposition between protected rooms and the corridor marks the latter as unsafe and unsanitary. Thus, the corridor frames Jeffrey's meetings with the mannequins by foreshadowing danger and building suspense.

On the other hand, the corridor marks the mannequins as marginalized and racialized. Echoing images of surveillance and illegal movement, the corridor is liminality's locale and where *a certain type* of characters appears. As Victor Turner points out, liminal characters have "no status, insignia, secular clothing, rank, kinship position, nothing to demarcate them structurally from their fellows" (98). Deprived of individuality, the mannequins become knowable only through their racialization. For, dominant discourse marks liminal spaces as "the border between civilized and primitive space, the space inhabited by savages whom civilized men vanquish on every turn" (Razack 13). Locating the mannequins in an over-determined liminality thus both replicates contemporary images of refugees and elicits already available fictions of racialized bodies. In this sense, DeLillo leaves no option of imagining the mannequins as any other than of color. He underlines, for instance, that they are darker – "reddish brown" (24). This is also how Jeffrey sees them – "painted in dark washes" (133) and "rust-colored" (51). Moreover, they are described as hooded, wearing chadors or burqas, standing "in the heat and dust" (51–52). In other words, they carry the characteristics of Muslim women in a desert-like environment, once again summoning racialized imagery.

Furthermore, DeLillo creates a juxtaposition between the white characters and white hospice and the dark figures and unsafe space. Thus, the figures contrast only one among them – the albino. Discovering a crypt in the Convergence, Jeffrey finds the mannequins:

> Here were figures submerged in a pit, mannequins in convoluted mass, naked, arms jutting, heads horribly twisted, bare skulls, an entanglement of tumbled forms with jointed limbs and bodies, neutered humans, men and women stripped of identity, faces blank except for one unpigmented figure, albino, staring at me, pink eyes flashing. (134)

Just a moment earlier, Jeffrey has observed another contrast: "there was a floating white light and I needed to put a hand to my face when I drew near deflecting the glare." The chiaroscuro re-confirms the darkness of the dark. Overdetermining their distinction from whiteness, DeLillo hardly needs to spell out the Otherness of the mannequins. Interestingly, had the story portrayed a postracial or post-post-

colonial landscape as Ferris suggests, there would be no references to white characters. Yet, the text abounds in such characterizations (46, 63, 87, 147–49, 203) and
thus further racializes other(ed) figures. Having shown how Blackness is rendered
absent/abstract and the mannequins made to signify otherness in *Zero K*, I now
want to offer one possible reading of Jeffrey's encounter with them – a **[jumpcut]**
between Jeffrey as non- racialized hero (as he appears in literary criticism) and
Jeffrey as a function of white author/ity (as I argue here).

Déjà vu

When Jeffrey meets the mannequin, he becomes overwhelmed with desire to
touch it. It is a sexual desire – Jeffrey wants to touch the breast in particular,
and we are told that this is something he has previously done. Throughout the
text, Jeffrey is positioned as the one who grows and gains authority. As the bicycle
passage reveals, the story often returns us, the readers, to past scenes. For Jeffrey,
touching the mannequin seems to have happened before as well. Then, how would
he have looked this past time, and how does the image of Jeffrey touching the
breast materialize before us? Considering DeLillo's descriptions of the figure, of Jeffrey, and the tensions between them, let me **[jumpcut]** to the "once upon a time."
There and then, Jeffrey stands at a slave auction (like a distant relative of Captain
Cook). He looks, touches and defines the value of what he sees. As Saidiya Hartman
writes in *Scenes of Subjection*, the auction was one of the stages on which forms
and formalities of white authority dis/figured physically and imaginatively, and
through sight and touch, Black bodies:

> Ethel Dougherty stated that at slave sales women were forced to stand half-naked for hours
> while crowds of rough-drinking men bargained for them, examining their teeth, heads,
> hands, et cetera, at equal intervals to test their endurance. According to Edward
>
> Lycurgas, enslaved women "always looked so shame[d] and pitiful up on dat stand wid all
> dem men standin' dere looking at em wid what dey had on dey minds shinin' in they
> eyes." Shining in their eyes and expressed in "indecent proposals" and "disgusting questions,"
> according to Tabb Gross, was the power, acquired and enjoyed by the owner, to use slave
> women as he pleased. Millie Simpkins stated that before they were sold they had to take
> all their clothes off, although she refused to take hers off, and roll around and prove that
> they were physically fit and without broken bones or sores. Usually any reluctance or refusal
> to disrobe was met with the whip. When Mattie Gilmore's sister was sold, she was made to
> pull off her clothes. (38)

Zero K dis/figures Blackness in a similar way. It allows readers today, some hundreds of years after slave auctions took place in North America, to imagine

them in the same light and on similar terms. Put otherwise, the mannequin – the form DeLillo assigned to Blackness – appears once again in terms of bodily availability. The readers' perspective is again directed to breasts, heads, skulls, and limbs. It is the perspective of a white male gaze, a perspective regulated by male desires and toward the figure's nakedness. In this sense, Jeffrey's vision aligns with what slave owners saw. He perceives the mannequins as "half bodies … stripped of identity" in their "faint yearning … the illusion of humanoid aspiration" (133–34). As Hartman argues, the slave system and its operative imagination restaged Black bodies as desirable and deathly things – as unhuman and calculable. What Jeffrey sees redoubles this imagination and it is once more redoubled the moment the story is read, because it is narrated with Jeffrey as the focalizer. Thus, DeLillo returns us to a scene of dehumanization by making it an available entry and marking it as implicit and logical. He renders anti-Black images usable and unequivocal, but also unique as each new reader approaches the text afresh in the claim of their own originality and abilities to interpret it. From this perspective, *Zero K* can hardly support the image and inspiration of historically conscious, subversive fiction, nor can it distance DeLillo as author/ity from white ideology.

Darkrooms

The above close readings reveal that DeLillo borrows from a complex and long-standing imagery: the opposition of ever-progressing whiteness and deadly/deathly Blackness. Black critique has long argued that this opposition is part of white American imagination, and at the heart of the story – the national myth – white America tells itself (see Morrison, Coates). The reason I zoomed in on DeLillo is not to single him out and thus threat as spectacular. After all, DeLillo is one in a long lineage of American writers who practice and capitalize on what Jess Row calls *white writing* (82). DeLillo's work might be central or paradigmatic in this category, but even so should not be isolated and treated as incidental. Writerly and readerly imagination often fails, as Morrison says, it "sabotages itself, locks its own gates, pollutes its vision" (*Playing* xiii) because it is not immune to ideology and power. I have taken *Zero K* as an example with the understanding that it, like any work of a white writer, will bear and bury its indebtedness to whiteness. I knew this before I read the book – as unscientific as this claim might sound, it best explains my argument.

Let us think of *Zero K* as an image developed in a darkroom. Before we have seen the final picture, or read the book, there is nothing on the photographic paper. The image, however, is already there; it is already taken – even if we cannot see it yet. With the help of certain techniques and technologies (an enlarger projects light and controls its intensity and duration), we have learnt to derive an image, one that can be seen and understood as such. Similar process happens in reading: language and literary devices all help in creating the appearance of a coherent whole. Because the image is there before we have seen it, we might assume that as viewers we have no influence on its structure; we have simply followed procedure and let it emerge. This is true only to an extent – lines and shadows appear on the paper, but it is our imaginative capacity to see them in context that allows for the assemblage of an image (see Saussure). For some images, this process happens several times – they are developed not only in the privacy of individual perception but in public discourse, as cultural products. This is what happens during canonization.

For a book to enter the literary canon, it must be assembled as canon-worthy. A majority of critics and readers must see similar features in it, which then develop in the positive version of what was already on the paper. Of course, much of *Zero K*'s popularity might owe to the fact that DeLillo is by now a well-established writer. As one of the "seminal, foundational figures" of American literature (Row 22–23), DeLillo grants a certain lighting which frames his works; his last, and in many ways conclusive, novel is bound to share in reputational fame. However, the novel also triggers interest and critical acclaim for what it tells its readers (this is how DeLillo entered the canon in the first place). As I have already mentioned, *Zero K*'s pixels cohere to a certain image; its themes and characters are read in general agreement over meaning, purpose, and effect. It is, as most reviewers say, a visionary and moral tale.

Ironical or not, *Zero K*'s representations are taken to be a creative but also ethical and political gesture. In the darkroom of literary criticism, the shapes of half-human mannequins, claustrophobic spaces where Blackness once lived, and abstract dangers eating up the world, combine as lines and form the image of post-race, post-identity, and DeLillo's own post- ideological practice. It is critical reception that processes DeLillo's text – a text that was there before we knew what to make of it and which was explicitly structured on racial and gendered divisions. Assembling individually and in agreement, critics understood this text as *something in particular*, and only then went on to re-publish, grant praises, anthologize, etc. The point I want to make is this: the *something in particular* can be bluntly racist, it can borrow and repeat stereotypes, and openly mirror ideology. Yet, reading practices, the processing of images which lead to understanding, always already move the lines. The focus falls on white author/ity to reinvent because the available devices and techniques are created to enlarge that image, to intensify light there – the positive version of a negative base, the recognition we can deal with and have dealt with for a bloody long time: *white men as creators, white men leading to white futures* (see *Futurity*).

Overlooking race issues in DeLillo's work does not stem simply from their marginal position. There is a structure in place that allows us to brush past them, to missee them even when they are in plain sight and central to the plot. At the end of *White Flights*, Jess Row asks: why are we still surprised? He notes:

> because whiteness has happened, the worst has already happened. It is not yet-to-come. The question isn't: why does the death exuberance of racism, of white supremacy, of necropolitics, keep coming back? The question is: why do we keep being surprised?

(286)

The answer to this might be unbearably simple. Despite all its insidious manoeuvres, anti-Black violence has always been before our eyes. As Martinot and Sexton say, it is "banal," "dissolved in the quotidian as aura," a "standard operating (police) procedure" (174, 177). The reason why anti-Black violence sometimes surprises white audiences as some unwelcome epiphany is because we are trained to missee it, to interpret the world past and through it. Whether DeLillo has crafted a racist product is beyond the point. Products like *Zero K* would expire if the system did not signify their self-sustaining and positive value, if looking-away was not already a part of white vision.

2 Reaching the limit, or how Kathy Acker used blackness to abandon Haiti and arrive home safely

Overview

Kathy Acker has contributed to reconceptualizations of female subjectivity both through her work and personal life. Many critics praise Acker for creating a new positionality for women and suggest that she underwrote it with originality and freedom. The risks Acker takes to arrive at this promising place, however, are complex and embedded in failure. Sometimes, Acker points self-reflexively towards it as we can see in her theorizations of abortion and piracy. Conceptualizing the latter as subversive and quintessentially feminist practices, Acker uses them to create a new world for women – by aborting the old and setting out to find a new one. In this sense, abortion has more to do with women's rebirth and renewal than with experiencing or inflicting death, as do the dangers of piracy.

Both abortion and piracy lead Acker's female protagonists to a new world; both involve some sort of distancing from the old ways of being, a sort of ridding themselves of tissue and traditional places/positionalities. Although piracy and abortion are intimately connected to death and danger, Acker uses them to imagine a renewed, better world. By framing piracy and abortion as women's transformative potential, Acker links them to the ways female bodies can move forward, and reach a brighter future. Before I can address my actual interest – namely, whether and how Acker racialized renewal – I want to contextualize this renewal. Piracy and abortion might seem an odd entry into this question, but they highlight some important aspects like renewal's limit and cost.

The limit and cost of the quest are often considered minor issues. How relevant can they be when the aim was a better world? Although sometimes disputed, Acker's feminism seems to be concerned precisely with this – the new, better place for women (Henderson, Pitchford). Would it matter if her means to reach this place were questionable, or some of her journeys towards it failed? One reason I focus on the cost and limit of Acker's conceptualization of renewal is that in the process of calculating risk and what can be sacrificed, it becomes clear who is deemed ex-

Note: An earlier version of this chapter was presented at the *Research Seminar Series*, School of the Arts and Media (UNSW, Australia) in October 2019; Excerpts of this chapter were published in *The Minor on the Move. Doing Cosmopolitanisms.*

https://doi.org/10.1515/9783110799996-003

pendable, easy to write off, to throw away. That what is expendable, the "waste" in the process, also defines the renewal for which it was disposed of. In other words, the renewal Acker reaches is not constituted merely by her wins (against patriarchy and heteronormativity, for instance) but also by her losses, sacrifices, and failures (i.e. "the waste product" of her struggles).

Acker wrote self-reflexively about the price of getting rid of things (time, tradition, energy, tissue, comfort, security, etc.) This becomes clear if we read her contemplations on piracy and abortion, and relate them to their purpose and effect in Acker's stories. Before I move to the question of renewal, I consider it important to see how Acker thematizes the price and limits of her struggle. Here, I am little concerned with witch-hunting Acker. Instead, I want to highlight that positive images, like the renewed place for women, come with "waste(d) material" and its production was neither wholly accidental nor entirely unconscious. I want to keep this consideration in mind when I turn to the question of renewal. For, the idealization of renewal's image (both in Acker's fiction and in her critical reception) often occurred alongside ignoring what is wasted, sacrificed, and failed.

I illustrate the above points by analysing Acker's 1978 novel *Kathy Goes to Haiti*. The text and its reception make clear how Acker's quest for renewal is idealized as emancipatory despite that Acker first uses and then jettisons Blackness (i.e. renders Blackness [waste] material). Although the novel is written as a parody of white women's imperialism, the main character cannot be read as parody because the white woman transcends the story: she is cleared of her exposed and mocked "dirt" and interpreted as inspirational, as a testimony of Acker's good intentions and success. This interpretation happens at the expense of Black bodies in the story, and outside the consideration that Acker's renewal might not be as inclusive and original as critics frame it.

Is it possible that Acker herself contributed to the quick "cleansing" of her white protagonist – for instance, by highlighting the greater good in dangerous experiences like piracy and abortion, and by framing pain and sacrifice as "just means"? But whose pain and sacrifice are we talking about? Moreover, is the greater good not subjective to the same principles that define which sacrifices and failures are ignorable? In the following, I will show how Acker uses racist imagery to write a seductive story (**kathy fails**), how critics ignore this racism (**failing kathy**), and how the image of transgressive white women conceals its racist corners for the sake of a brighter object (**kathy the pirate**). In all these moments, whiteness plays a crucial role.

The novel, parody, and movement

Sometime in the 1970s, the-now rebel icon of queer punk literature Kathy Acker was commissioned to write a short porn novel and travelled to Haiti to do her research for it (Dandicat vii-viii). The book that came out of this, *Kathy Goes to Haiti*, is one of Acker's least analysed works and despite her oeuvre's general focus on sex and sexuality – one of the few actually pornographic stories Acker came to write. In their 1986 conversation, Acker corrects Angela McRobbie's assumption that most of her texts were written as pornography:

> I did to some extent write [*Kathy Goes to Haiti*] as a porn book, I just wanted to see whether I can write a porn book, that's the only time I did that…and even then I was cutting away … you know, just to take the piss out of it. ("Interview with McRobbie")

Acker explains further that, albeit relative and contested (she was still taking "the piss out of" the genre), the main difference between *Kathy Goes to Haiti* and other depictions of sex in her work is the de(con)struction intended by the latter. Pornographies, in contrast, "stimulate erotic … feelings"; pornographic illustrations are explicit and exciting and aim to build desire up (*OED*). The distinction Acker draws in the conversation with McRobbie is not one of language (after all, explicit sex is Acker's insignia) but one of intent and effect. *Kathy Goes to Haiti* aims to arouse and satisfy its readers – literally, as pornography, and only tangentially – as an Acker's novel, one that takes the piss out of desire and works through its double entendres and degrees.

Kathy Goes to Haiti dissatisfied Acker. She dismissed it as "dumb" and "conventional," and often reminded audiences that she wrote the text for money and that "nothing happens" in it. *Kathy Goes to Haiti* might be indeed Acker's "most linear and conventionally narrated" story but its significance has gone underappreciated by Acker and her critics ("Interviewed by Deaton"). This misrecognition can certainly be attributed to the text's conventionality – lacking the typical for Acker experimentation, the novel at least to an extent reflects rather than deconstructs dominant discourse. I say to an extent because Acker wrote *Kathy Goes to Haiti* as a parody of porn novels and Nancy Drew stories and, as Clay Guinn maintains, of white female tourism and imperial instincts ("You Get Off").

Linda Hutcheon clarifies that postmodernist parody is a "value-problematizing, de-naturalizing form of acknowledging the history (and through irony, the politics) of representations" (90) and Acker's recourses to parody have been often read as such denaturalization (see Colby, Borowska). Hutcheon notes further that feminist postmodernist parody, to which category most literary critics assign *Kathy Goes to Haiti*, often presents "new kinds of female pleasure, new articulations of female desire [and] by offering tactics for deconstruction [subverts] the patriarchal visual traditions" (156). From this perspective, the least *Kathy Goes to Haiti* promises is to be less straightforward than Acker claimed. On the contrary, as parody the novel incorporates and ironizes its source materials – race and sexuality as this chapter establishes – and is thus a fitting place to examine the way Acker diagnosed and attempted to deconstruct them.

Paradoxically, the conventionality of the novel might be exactly what makes it so important for literary analysis. Commissioned, the book had to appeal to larger audiences and as pornography, to arouse and satisfy much more directly than Acker's other texts. Acker's lack of interest in *Kathy Goes to Haiti* can be linked to this accessibility. She explains:

> I took the formula of a porn novel and I made a structure, a mathematical structure. In this chapter we have psychology, and here we have this happen... and then I wrote it according to this structure. It was the most boring thing in the world to write. I tried to make the characters as dumb as possible and I basically tried to make nothing happen. ("Interviewed by Deaton")

The "compromises" Acker settled for – partly to prove that she can write a conventional text, and partly for the eight-hundred dollars porn publishers were giving for one – respond to the specific literary scene and expectations the novel was tailored to fit (Acker, "A Conversation"). *Kathy Goes to Haiti* contains and comments on these expectations, and is also indicative of what Acker kept and aborted from her style in order to address them. As such, the novel offers a glimpse into Acker's negotiations not only between mainstream fiction and Acker's praised avant-gardism, but also between her de/constructions of sex and sexuality and their articulation as "sexy," as suitable for porn press and readers.

In my analysis, *Kathy Goes to Haiti* yields precisely the investigation into what Acker was ready to abandon and risk in order to arrive at a desired outcome. To me, the book indexes Acker's narrative manoeuvres in pleasing and parodying dominant discourse but it also reveals its bare structure – the basic parts of the story, the parts Acker could not go without, and as I will argue, organized the readability of *Kathy Goes to Haiti* and its subsequent celebrations as a good, and fundamentally emancipatory story.

Crucial for including this novel in the section *Movement* is Acker's use of tropes like mobility and stasis in sustaining narrative cohesion, and for representing/parodying ideology. Interestingly, *Kathy Goes to Haiti* shifts between pornography and its parody, and mainstream and experimental fiction, and in transitioning from these frameworks marks a new, negotiated place between them. Moreover, the novel offers an occasion to examine how Acker's reputation as an experimental, politically conscious and leftist author transfers onto its reception despite the fact that 1) Acker considered *Kathy Goes to Haiti* a minor and ineffective work, and 2) the text fails to deconstruct dominant ideology and exposes Acker as complicit in rather than critical of it. As in the section on DeLillo, I build my analysis around ideas of movement – as something Acker's protagonist does to arrive to a new place, as something that occurs when Acker's reputation influences interpretations of *Kathy Goes to Haiti,* and as a way images of whiteness infiltrate reading practices at large.

Focusing on Acker's conceptualizations of failure and the way strategies and devices malfunction in her texts, I consider whether and how literary criticism brushed past the possibility that Acker, "the pirate-queen of avant-fiction" (Olsen, "Kathy Acker: Queen"), might have not arrived at as liberating a position as she is often said to embody. Examined in this light, movement refers to the imaginative shortcut interpretations overlooking failure allow, and explains why misrecognitions of this failure have kept Acker true to her status as a subversive hero. More importantly, movement in the sense of imaginative flight from undesired insight/information (in this case *Kathy Goes to Haiti*'s anti-Blackness) highlights which of Acker's transgressions and failures were considered ignorable and did not bruise her reputation.

Kathy the pirate

> I'm no longer a child and I still want to be, to live with the pirates.
> Because I want to live forever in wonder ... it matters little whether
> I travel by plane, by rowboat, or by book. Or, by dream. I do not see,
> for there is no I to see. That is what the pirates know. There is only
> seeing and, in order to go to see, one must be a pirate.
>
> Acker, *Bodies* 158

It would be fair to memorialize Acker as Kathy the Pirate. Both her personal life – she was a queer, sex-positive deserter from upper class family and formalities – and her work – were a continuous experiment with language, body, and consciousness[1] – promoted the image (see Colby, Pitchford). Braced by her appropriation technique (Acker preferred to call it *piracy*) and her recurrent characterisations of women as pirates, the figure of the rebel venturing in uncharted territories has come to define Acker as a hero of the margins in pursuit of adventure and power. Indeed, Acker attacked heteronormative ideologies, eagerly crossed borders and traditions, and desired moving, transformative texts (see Sciolino). She also

---

1 think of Acker's experiments with Dream writing (see Pyke, Colby, Kraus)

found herself at the edges – of herself and society. As Ellen Friedman notes in *Kathy Acker and Transnationalism*, Acker embodied risk and dissent:

> Cathy, Kathy, the shapeshifting personae, despised, isolated, raped, beaten, robbed, abandoned, lonely, humiliated, with hundreds of stories, poems, histories, philosophies streaming through her texts – repeatedly reemerging in yet another underworld as picaresque pirate, criminal, prostitute, zombie, a reviled wandering Jew carrying the books of the world in her head. In her cut up technique, they come out of her fractured, vomited back at the prevailing culture, reformed and accusing. (xiv-xv)

Lanse Olsen calls Acker "arguably the most important postwar pirate-queen of avant-fiction" ("Kathy Acker: Queen") and this image continues to attract attention to her as simultaneously marginalized (avant-garde and pirate) and cult and canonized (the most important queen). For Georgina Colby, Acker continues radical modernism and the lineage of innovative and eccentric writers like Gertrude Stein, James Joyce, H.D., and Ezra Pound (see *Writing*). Described as punk feminism, Acker's eccentricity and innovation were often deemed political and unpopular (until they were not).[2] As Chris Kraus reminds us, Acker quickly turned to a cult figure, a superstar (*After*). This is, in fact, how she was billed for a performance at the Australian Centre for Contemporary Art in 1995: "Kathy Acker US Superstar Punk Feminist Writer" (260). Acker often embodied oxymoronic characterizations like "superstar"[3] and "punk,"[4] and even descriptions like "feminist" sometimes seemed autoimmune in her. In her defiance of fellow feminists, Acker proved disloyal, determined, and one-of-a-kind (see Pitchford). Her habitual straying away from fixed categories and movements further braced Acker's artistic personae as the pirate, the rebel.

Ascribed to Acker, piracy cannot be limited to image(s). In fact, she continuously reflected, theorized, and manipulated the image to extend agency and criticism. Through piracy, Acker exposed past oppressions and spoke back to and against canonised texts and celebrated writers. She considered many of them (often inad-

---

2 According to Peter Wollen, Acker gained commercial success around 1984 with the publication of *Blood and Guts in Highschool* (1–12); Acker's success was called into question around 1989 (Kraus 232–244)

3 "an exceptionally outstanding performer in the theatre, music, sport, or some other public sphere; a superior star" and "an exceptionally successful, advanced, etc., thing." (*OED*)

4 "(a) A person of no account; a despicable or contemptible person; (broadly) a person, a fellow (rare) and (b) a petty criminal; a hoodlum, a thug.; U.S. slang. **a.** An amateur; an apprentice; "A young person, or a person regarded as inexperienced or raw; Contemptible, despicable; thuggish; cowardly; inexperienced, raw.; Of poor quality; devoid of worth or sense; poor, second-rate, inferior; "lousy." (*OED*)

equate) teachers and heroes, and referenced their works by pillaging entire passages, styles, and ideas. Thus, Acker's piracy made language and ideology more visible (see Sweet, Colby). It also foregrounded Acker as someone with a place in history. In fact, piracy proved that Acker came after those she took from and that she anticipated her own re-emergence, that literature was moving, and that the successes and failures of her predecessors are forever embedded in her writing:

> Only the incredible egoism that resulted from a belief in phallic centrism could have come up with the notion of creativity. Of course, a woman is a muse. If she were the maker instead of the muse and opened her mouth, she would blast the notion of poetic creativity apart. ("Few Notes" 33–34)

Surely the above statement can be read as postmodern scepticism and nullification of originality. To me, however, it marks Acker's indebtedness to writerly influences, to the symbiotic relationship between the "maker" and the "muse," where creativity flows and ties the two and disappears (at least in the sense of individual property). In a similar vein, Martina Sciolino suggests that Acker construes her personae as a "maker who is herself made" (440). Logically then, intertextuality and Acker's literary interlocutors appear straightforwardly, almost banally in her texts – she borrows writers' words, ideas, and also uses historical figures as characters in her fiction. Pirated and bygone, Acker's muses surface with the risks they took and those they did not. In this sense, piracy reveals Acker's desire to transgress but also highlights the self-reflexivity that she is herself a product of past transgressions.

This connection materializes clearly in Acker's decision to borrow from various texts and cultural producers. As Paige Sweet suggests, the manoeuvre points not so much to Acker's plagiarism but to her resolve to communicate and dissent. Furthermore, piracy describes the ways Acker understood property and ownership, and exposes the narratives that sustain them:

> Piracy ... challenges the legal categories that protect and even sanction one kind of thievery (that which operates on behalf of capitalist accumulation) while criminalizing another (such as copyright infringement). One might thus read [Acker's] piracy as a kind of taking back, a reclaiming of a previously stolen good. But it is not merely a restorative gesture ... Acker renders explicit the violence, perversity, and economic inequalities that are implicit in the texts she pirates. In this way she produces a literary event that is able to illuminate the injustices of the historical and political circumstances that inspire much of her fiction. ("Where's the Booty")

I follow Sweet in examining piracy in light of its critical potential. I would like to underscore, however, that this potential arises with Acker's reclaiming of piracy as a feminist practice. She attests, for instance, that "the separation between [her] and piracy had something to do with being a girl. With gender" ("Borrowing"). Indeed, piracy is marked as subversive, but in Acker's work it remains quintessentially a woman's question. Configured as something only men dare to do, piracy requires capacities patriarchy refuses to see in women: the faculty to risk, break free, and navigate their destiny. For Acker, piracy was women's chance to rule; their new role was wrestling with the body and its right to offend, wound, and violate. In fact, wrestling (with the female body) was a central concern for Acker, and she often reclaimed male-dominated spaces (like the gym), and male-dominated practices (like exercising strength). Neither of these ventures guaranteed success:

> What actually takes place when I bodybuild? ... Bodybuilding can be seen to be about nothing but failure. A bodybuilder is always working around failure ... By trying to control, to shape, my body through the calculated tools and methods of bodybuilding, and time and again, in following these methods, failing to do so, I am able to meet that which cannot be finally controlled and known: the body. ("Against Ordinary Language")

For Acker, coming to terms with the body involves recognition of its ability to fail, to be exposed to danger and lose (or get lost). Like bodybuilding, piracy works to "shock [woman's] body into growth" but against idealization as the only way to reclaim female positionality ("The Language"). On the contrary, a woman's body (of work) need not be perfect to resist and survive. Like bodybuilding, piracy reveals that power and agency involve loss, fatality, and failure. The struggle to see might involve sacrificing the body – "there is no I to see" as Acker writes in *Bodies* – but one must go and see nonetheless.

This becomes clear if we examine how Acker fuses piracy as a form of protest with women's right to mis/carry unwanted traditions. Relating the insides of piracy and abortion, Acker underscores female courage (both practices require it) but also suggests that there is a price for risking one's own body, for reaching too far, or not far enough. For, to abort or to venture into unfamiliar waters can turn equally deadly. Abortion and piracy both figure as a moment to re-discover one's own capacity to choose, and inevitably – to find out what one is capable of, what one desires. Piracy and abortion also signal danger, (self)sabotage, and failure. Acker never avoided acknowledging failure, nor did she avert the question how far minor bodies can reach when danger was looming. Sometimes the scariest thing in unfamiliar waters was the water itself – a mother's risk, the water breaking, a child on the way. As Dodie Bellamy suggests, this is the story Acker consistently returns to:

> Over and over, Acker tells the same tale: the mother is pregnant with the daughter, and the father leaves. The mother blames the daughter and tries to abort her. The daughter's body survives, but not her unified self. (qtd. in Kraus 14)

As with piracy, Acker manipulated the meaning and images of abortion to articulate the complexities of desire and power. By exploiting their hydraulics and metaphorics, Acker connected the singular and the spectacular – piracy – to the quotidian, to abortion as an experience familiar to many women and indicative of their capacity to endure transformation.

Interestingly, the practice of piracy mimics that of abortion and situates the works Acker appropriated – let us take for example Dickens's *Great Expectations* – inside her texts. Thus, Acker's piracy allows for containing and owning someone else's invention. The latter appears foetus-like, internal, like something Acker had welcomed into her own body of work. Acker's 1982 rewriting owes much to Charles Dickens because he helped her "conceive" and his presence is visible in the first pages of Acker's novel. Dickens's originality is then not only ironized but also made to generate new creations, this time with Acker. She performed this conception many times, unashamed to do it with anybody. After Acker took from existing cultural productions, however, she aborted and revised them, and made them her own (see de Zwaan). Thus, Acker visualised a process of literary and literal creation, and marked the moment of woman's right to choose, to change the course of narrative direction, and continue otherwise despite the ever-present shadow of the source text. As Georgina Colby suggests, Acker's works

> offer alternative linguistic configurations that break away from inherited background and attempt, in doing so, to find new forms of knowledge that are not contingent upon a background bequeathed by tradition. (12)

From this perspective, Acker's piracy can be said to signal the conscious and rebellious refusal, the thinking-away from what is given and prescribed. Paired with the conceptualization of abortion, piracy becomes something more than a simple venture and theft. For, in taking someone else's work and then aborting it, the pirate queen blasts ideas of creation and property but not before pushing her body to the forefront. Yet, the refusal (of a child or tradition) is inevitably marked by culpability. As we will see, Acker's decision to mis/carry legacy and tradition prove troublesome, far beyond the idealization with which Acker is sometimes celebrated.

I understand experimentation and risk as something Acker intentionally worked on – fearfully, in pain. Revealingly, her narrative choices often bring forth women's journeys in dangerous territories. Piracy then remaps bravery and survival in their transhistorical articulations as Sweet suggests. It is one mode of Acker's social critique and power to object. It also illuminates, however, the specifics of woman's right and risk towards what is not yet born and imagined. In this light, piracy hints towards Acker's own concern and ventures in unfamiliar waters. Being a pirate, in fact, is not too different from taking risks from a feminist positionality and experimenting with one's own body *sometimes that's all we've got*.

Thought together, piracy and abortion map the possibility for women's misfortune, death, and errors. For instance, Janey, the protagonist in *Blood and Guts in High School*, explains:

> I wanted a permanent abortion ... Abortions are the symbol, the outer image, of sexual relations in this world. Describing my abortions is the only real way I can tell you about pain and fear. (33–34)

In "Requiem," the last work Acker submitted for publishing to *CTHEORY*, George tells Electra that the abortion had not worked because she is meant to be born. Much earlier, in 1983, when Mark Magill interviewed Acker for *BOMB Magazine* and asked why she was born, Acker simply responds that her "mother was scared to have an abortion." Although matter-of-fact, the reply is ambiguous and refuses to answer Magill's second part of the question: was her conception planned, accidental, or neither. Instead, Acker's response directs Magill to the choice after the conception. It thus associates Acker's life with her mother's fears rather than her intimate relations. Not only is the father removed in this manoeuvre, but conception is transferred onto the plane of women's power and will, not sex. Analogously, Acker concludes in *Bodies* that "women's freedom ... depends on her ability to stop pregnancy" (70).

Yet, none of Acker's texts posits female power and freedom as inexorably consistent, absolute, and triumphant. Acker's protagonist in *Don Quixote* confesses, "I've had a dead abortion ... I mean: an abortion by a horse. I need you to take care of me" (16). Here abortion is successful but inadequate, and imposes burden. Similarly, in *My Death My Life by Pier Paolo Pasolini*, Acker writes: "I keep trying to kill myself to be like my mother who killed herself" (222). This attempt is protracted, bloody, and painful. It endangers the female body. It also implies that women face risk openly, in full consciousness. On the one hand, Acker's protagonist tirelessly corrects her mother's failure to abort. For Acker, the failure to abort is a central question and she used it as a metaphor remapping women's guilt, complicity, and limitations. On the other hand, Acker found the im/perfect link between herself and her foremothers: the persistence of failure. In that sense, feminist lineages did not necessitate all-inclusive agreement with and idealization of other women. Instead, female subjectivity asked for catching the unwanted early and confronting the fact that even the most intimate and invisible risks can re-position women's bodies.

Acker's work explicates that these risks come at a price, do not triumph with immediacy, and demand sacrifice. Framed by piracy and abortion as subversive but noxious practices, Acker's courage to re-invent and create reveals how certain bodies bear experimentation and are driven to death. Re-reading *Kathy Goes to Haiti* from the perspective of Afropessimism shows, for instance, that Acker's attempt to miscarry certain traditions but remain desirable within discourse comes at the cost of Black bodies. Literary critics consistently overlook this cost. To an extent, the misrecognition might owe to Acker's reputation and her devotees' objective to keep this reputation functional. After all, pirate queens transgress, the transgressions only reconfirm their rebellious and courageous nature. A bloody, painful journey seems to be a given when the struggle is to remap the world and women's place in it. Reading Acker, I have often thought that this is a truly empowering, even heroic perspective (*"there is no hero without a wound"* goes one *Bulgarian saying*).

Informing my readings with Black critique, I realized that an important perspective is missing from the above conclusion. Namely, who bears the pain, whose blood is sacrificed? Who is thrown overboard when the journey becomes dangerous, when waters are risky? The quest for (new) land is not an inadequate historical analogy here **[jumpcut]** the ship Zong, 1781, captain Luke Collingwood makes navigational errors which lead to a protracted journey. Water supplies are running out. Zong is a slave ship – its British crew transports enslaved Africans from Accra to Jamaica. In order to save water and food supplies for the extended time at sea, the crew decides to throw ca. 140 enslaved Africans overboard. Much has been said about the Zong massacre and the subsequent court case, about the non/murder of people considered cargo, and the impossibility to fully grasp the violence exerted on their Black bodies and mourn their destruction until some sense of peace and redemption be found (see Philip, Baucom).

> Redemption is the narrative inheritance of Humans
>
> (Wilderson, *Afropessimism* 325)

At this point, I want to consider this – the crew's desire to reach safe land, and the always already calculated decision of who is expendable, who can be wasted (along the way). Black critique teaches us, that this calculation cannot disappear easily, that, in fact, it has not disappeared to this day even if the label of the Black body has changed from "slave" to "free" (Hartman, *Scenes* 25). The logic behind the calculation stayed (see Wilderson, Sexton, Sharpe) and so, in dominant discourse, Black bodies continue to be framed as "available equipment" that can be used and thrown away (Warren, *Ontological* 76).

> Objects exist as implements, tools, in the psychic life of Human subjects
>
> (Wilderson, *Afropessimism* 198)

The manner of relieving white desire by wasting Black people as material is visible today despite the considerable **[jumpcut]** one needs to make between a pirate queen like Acker and a British ship crew.

Why have Acker's uses of Blackness in *Kathy Goes to Haiti* remain ignored? One explanation is that practices of theft and accumulation are already in place, and sanctioned by dominant discourse (see Morrison, Wilderson, Hartman). These practices are longstanding and pervasive – with their logic seeping from one event to the other, often remaining unobstructed because of its constant and invisible movement (*Racecraft* 18–19). If we look at *Kathy Goes to Haiti*, we will see some of the effects of such logic. We will see, for instance, how Kathy Acker renders Blackness useful to her literary project, how she utilizes it as *fungible* (to use Hartman's term) in struggling to find a new and better place for white women. Thus, Acker extends narrative coherence through the legitimacy already granted to cultural appropriation and symbolical colonization – practices underlaying white American literature as Black critique has shown. Toni Morrison explains,

> autonomy, authority, newness and difference, absolute power [are not only] the major themes and presumptions of American literature, but … each one is made possible by, shaped by, activated by a complex awareness and employment of a constituted Africanism. It [is] this Africanism, deployed as rawness and savagery, that [provides] the staging ground and arena for the elaboration of the quintessential American identity (*Playing* 44).

In the following, I will show that *Kathy Goes to Haiti* does not bury such construction. On the contrary, the white protagonists' progress shines in obvious juxtaposition to dehumanized Black characters; the opposition charges the text in blunt, straightforward ways. To build the narrative, Acker relies not only on Kathy (her/self) as a representation of American identity but on the relationship she establishes to Haitian Blackness, and the ways she sacrifices it in her journey. In this sense, Acker takes from an available imaginary deposit, from Blackness open to her narrative transgressions.

This is neither a new, nor original practice. As Fred Moten remarks, white imagination – spanning from the first transgressions of colonialism to the latest execution of anti-Black violence – splits Blackness open so that "anybody can claim it" ("Do Black Lives Matter"). In a similar vein, Wilderson notes that whiteness positions "the Black in an infinite and indeterminately horrifying and open vulnerability" (Red, White, and Black 38). Bringing Hortense Spillers' theorizations on Black destruction as a "living laboratory" together with Saidiya Hartman's analysis of the fungibility of Blackness, Calvin Warren concludes:

> Black ~~beings~~ constitute this irresistible source of availableness for the world ... Instruments, tools, and equipment are interchangeable/replaceable; this is starkly different from human being, whose existential journey in the world renders it incalculable and unique ... black ~~being~~ is pure function or utility. (*Ontological* 46–7)

Kathy Goes to Haiti is obvious and immediate in exploiting and jettisoning Blackness. In using Black characters as a background against which Kathy shines as a hero, Acker utilizes derogation to organize the narrative and, essentially, the reconfiguration of Kathy – her own image. Furthermore, Black characters enable interpretations of *Kathy Goes to Haiti* as a political text. In this, Blackness performs a double function with/in Acker's narrativization. On the one hand, it affords readerly entertainment (fetishization working for both pornography and parody). On the other, Black characters lend what can be called an Acker-ian twist to the novel – their presence allows critics to read the text as social critique, as part and parcel of Acker's reflections on history and politics. Yet, no attention has been paid to how 1) a derogation of Blackness has served to validate Acker's arrival at a new, better place, or how 2) an experiment of language extends, rather than escapes, the source text, the grand narrative from which Acker is said to depart.

Literary critics have argued that Acker satirises the imperialistic impulses of her foremothers and thus claims a more ethical position for white women (see Riley, Guinn). As I will show, this claim is made possible with the connections Acker drew between Blackness and white female renewal. Before discussing the failure of Acker's attempt to abort racism, however, it is necessary to consider why Acker might have been exonerated from the act of failing in the first place. Having addressed the functionality and some crucial aspects of Acker's reputation, I now focus on the meanings drawn from it and transferred onto *Kathy Goes to Haiti*. Allowing Acker to appear as deadly and dying, as truly a Pirate, my study departs from admirations of cult personas, of Acker the idol, and instead exposes her to failure and critique, but also to the transgressions achieved by, and for, Kathy, Acker's own image.

Failing Kathy

Acker is frequently celebrated as the hero who dares to live on the edge and fight against oppressive normativity. As Katie Mills notes,

> Kathy Acker brings to avant-garde literature the angry gender politics [and her] novels sketch out the goal of female protagonists to find a voice that will "transport" them to new places. (180)

Moreover, Acker's capacity to risk and disobey is often described as restorative. Like Sweet and Friedman, Ralph Clare understands her work as social criticism and underscores its links to originality and renewal:

> In a world in which liberal democracy is itself in crisis, Acker's work is more relevant than ever. Change the political genre, fight for something new, Acker's work urges us, because conventional politics in the post-factual, image-driven age is failing and unable to evolve. ("Why Kathy")

Clare fixes newness to Acker's break with conventional politics. Like most critics, he describes Acker's recourses to change and transformation as a regular, anticipated direction, as an "endless process, not an end product." Interpretations of *Kathy Goes to Haiti* follow a similar vein. Shannon Riley and Clay Guinn have commended it for deconstructing the travelogue as a tool of hegemony and as a critique of imperial instincts. Furthermore, Guinn argues that Acker's parody works against the longue durée of dehumanizing Blackness. Alongside this appraisal, Guinn references the feminist Sara Mills:

> Sara Mills, for example, suggests that the female travel writers exist outside of the traditional imperial relationship. Female writers, Mills argues, "cannot be said to speak from outside colonial discourse, but their relation to the dominant discourse is problematic because of its conflict with the discourses of "femininity," which were operating on them in equal, and sometimes stronger, measure" ... These are important voices in postcolonial scholarship, but I do not find them applicable here. While it is tempting to think of Acker in terms of a female voice, she was often sceptical of feminist categories ... and often sought to subvert them. In *Kathy Goes to Haiti*, the heroine is obviously female, but I believe that she is more a representative of American hegemony than she is restrained by it. (17)

Guinn relates Acker to feminist reconstructions of history but implies that she surpasses their deficiencies (or, that Acker overcomes Mills' exoneration of white women from anti-Black impulse and violence, and Mills' and the heroine's racism). The novel is seen as a corrective of feminist failures, or at least as their timely reprimand – white women also exoticized Haiti and exercised power, and they did it in real (literal), and literary, movements. As Guinn concludes, Acker has established that "stereotypes ... are the product of our collective cultural imagination" (15). In a certain way, Guinn charges Acker's acknowledgement of feminism's errors with the implication that the capacity to recognise failure helps Acker escape rather than simply reflect on it.

On the one hand, Guinn's account suggests that Acker has overcome herself by criticising white women's sexual exploitations of Haiti, that is, by exposing women's participation in imperial projects. As a personification of white female imperialism, Kathy's character enables Acker to critique imperialism, to expose and ironize it. On the other, Guinn frames Acker's protest as a collapse of identity – Kathy figures not so much as a female character but as a representation of American hegemony (17). In a similar vein, Georgina Colby notes that Acker's intertextual writing "enables the resistance to fixed identities and a singular self within the text" (142). For critics, Acker's protest often surpasses identity politics and extends to structural matters (see Guinn, Colby, Kocela). In this light, opposition to hegemony appears to succeed through body-less prowess because once the character signifies risk and revolution, she is no longer constrained to bodily singularity and individualism. She can disown and occupy any role and be equally transnational and local, marginalized and canonized, sexually free and sexually oppressed, fictional and historical. In a similar vein, Clare maintains that:

> There is a politics to Acker's fiction that goes beyond her critique of identity ... Acker seeks a "freedom" prior to those "freedoms" guaranteed by, say, the state, a constitution, or the law ... She counters it with a politics that ties the body together with a notion of a freedom ... freedom as an endless process, not an end product. ("Why Kathy")

The implications of readings like Guinn and Clare's are significant. If one examines *Kathy Goes to Haiti* informed by such analysis, they could easily conclude that while Kathy travels to Haiti in pursuit of sexual freedom, Acker's parody of the journey manages to exceed white feminism's failings. Such reading would infer that the revolutionary woman/writer has skipped time and place to return to a moment of freedom prior to both, and that Acker's text has successfully become a feminist critique against women as "makers," and against women as "made" – as part of the system that produced them. In this sense, Acker's critique seems to overcome the past but also to move forward – renewing and safeguarding the future.

The idealization of Acker's social criticism erases her self-reflexive writing on women's precarity and struggle. It also occludes Acker's failure to abandon racism. As I have argued previously, Acker's piracy and thoughts on abortion are indicative of the limitations Acker saw as inherent to risk and revolution. Piracy and abortion point towards a kind of understood impossibility of completely filling that-which-has-been-taken-out, be this child or tradition. In this sense, piracy and abortion reveal Acker's insistence on recognizing the female body (of work) as capable of surviving, prospering, erring, and dying. Acker also defended this in her personal life, to her death. Perhaps most clarifying is her refusal to treat cancer in the safety of orthodox Western medicine and instead keep "doing what she wanted, when she wanted, how she wanted" (Mcbride, "The Last Days"). Her (auto)biographies repeatedly expose the anger and disapproval of Acker's friends and family and their refusal to accept her wrestling with the decease on her terms. She disagreed and risked her body in the ways she saw fit. In doing so, Acker inadvertently criticized demands on female functionality and endurance. She also welcomed risk and the potentiality of failure in her own body. Acker confesses:

> As I walked out of the surgeon's office, I realised that if I remained in the hands of mainstream medicine I would soon be dead, rather than diseased, meat. For conventional cancer treatment was reducing me, quickly, to a body that was only material, to a body without hope and so without will ... My search for a way to defeat cancer now became a search for life and death that were meaningful. Not for the life presented by conventional medicine, a life in which one's meaning or self was totally dependent upon the words and actions of another person, even of a doctor. I had already learned one thing, though I didn't at the time know it: that I live as I believe, that belief is equal to the body. ("The Gift of Decease")

Acker did not beat cancer. She had followed alternative routes, she had also proved to the world that there is nothing immune and holy in her. This is not the only time she failed. I suggest, however, that friends of her body (of work) like Guinn and Clare, and those from her community, also fail Acker. They do so in the insistence that she keeps her body wholesome and flawless, as if it could take on all the cancers in the world, and should inevitably triumph. In that, even well-meaning supporters of Acker follow the longstanding traditions of 1) oppressing women by demanding their body's pure functionality, and 2) overlooking certain failures, deeming them ignorable, in order to protect risen idols, and keep the record clean.

Dying at a Tijuana clinic for alternative medicine, and having left many friends and family behind, in 1997 Acker seemed to communicate a different story – one in which the aspiration to fight against social or bodily cancers does not guarantee recuperation. Instead, this ending of the story shows that women carry wounds and the capacity to wound inside their guts (*guts as a triple entendre: stomach, courage, essential part; and as a gesture towards Acker's legacy: her, first forbidden, then idolized, Blood and Guts, the experiment, the risk to disobey, the pirate in motion*). Acker's story ends with her attempting to escape but ultimately, and intimately, losing to her own image, to Kathy in writing, the one who outlives in re- readings and reaches only as far as we are ready to take her:

> I could've been Kathy. Kathy could've been me. I don't know. I could've been you, you could've been me. We all could've been Eleanor Antin. It's all the same. And by that I don't mean we're not who we are. But you know what I mean.
> (Rosler, qtd. in Kraus 279)

This is what the artist Martha Rosler tells Chris Kraus when they talk about Acker. Why would Rosler conflate identities with the above statement? What is it that makes certain figures relatable, even substitutable? Can we say that Rosler, Kraus, Acker and Antin are related in their Americanness and their art, or maybe in both? If yes, does Acker disappear or re-emerge in such relation, i. e. does Rosler confirm that Acker is truly gone or precisely the opposite – that Acker is renewed in other images of herself?

I made a similar statement to Rosler's when I argued that a conflation of Acker's white protagonist, white reader, and Acker as author allows us to highlight, to see, how white authority works for them. Like Rosler, I was not interested in "who we are," not on an individual level at least. Rather, I wanted to understand how we – a white writer, a white protagonist, and a white reader – follow, use, and become author/ity to keep going, to gain a footing. *Texts are like oceans – if I can hear Èdouard Glissant from here.* By now we know that whiteness operates "as a strategy of authority" (Bhabha, "The White Stuff") and that this authority is not given but acquired by violating Black bodies (Yancy 237– 238). It matters little who takes up the task of exercising white authority – it is available and pervasive, and although it can change its actors and perfect its practices, at its base it remains a system for furthering white being by damaging Black being (Coates, *Between* 10). In the following, I look closer into white authority (as moving between Kathy, Acker, and white readers) and how it works to damage Blackness.

Kathy fails

> Your mother didn't want you, disliked you ...
> You kept looking for a home. Your need gathers ...
> You are in it now, baby
>
> (Haiti 76)

Having discussed how literary criticism positions Acker as a saviour from dominant discourse, I now turn to some of the ways her personification in *Kathy Goes to Haiti*, the character Kathy, is portrayed as a figure of "doubled" liberation despite her racist manoeuvres. As I will show, literary criticism takes Kathy's characterization to signify 1) Acker's courage to further and protect feminist agendas (Acker as "transformative and accusing" as Friedman suggests), and 2) Acker's vulnerability (Acker as "the raped, beaten, robbed, abandoned, lonely, humiliated," again pace Friedman) and thus women's need of salvage. I argue that these images of Kathy mobilize whiteness, and organised through the white saviour and damsel-in-distress tropes, lead her to a renewed, better place. A closer look at these tropes reveals how Kathy's movement ahead, her liberation as a subversive figure and a character in danger, is fuelled by a juxtaposition to a dehumanized and static Blackness.

The story is simple. Kathy, a young white American girl, leaves home to explore sexual pleasures in Haiti. There she meets many men – mostly Black and always lusting – and a few sickly and stuck in Haiti white women. Adventurous and wandering, Kathy tries to get where and what she wants. This proves difficult because Haitian men stall her, nothing happens in Haiti, and Kathy's own naïveté slows her down. Turbulent seem to be only Kathy's desires to move in, to, and away from Haiti, and her orgasms with Roger Mystere, one of the "mulatto robber barons" with whom Kathy falls in love (53). Bracketed by Kathy's arrival and her anticipated return to America, sex scenes turn on a loop and mirror the persistency with which Haitian men drive Kathy crazy, and in circles. At the end of her journey, Kathy returns to America having gained not only hedonistic enjoyment but also maturity and self-determination. [5] Reviewing *Kathy Goes to Haiti* on *goodreads*, one reader declares:

> This book made me feel like someone's mother. My goodness Kathy make me nervous about Haiti. I lived in Haiti, and I know that's a very very bad place to. . .hitchike, sleep with random people, wander to towns that you don't know. . .and the like. I half expected her to get hurt or killed. I suppose she did fine, but still–what a tense read for me! (*"Kathy Acker Goes to Haiti:* Review")

Protecting Kathy is not an entirely unsubstantiated desire if the novel is read as linear and straightforward. Although twenty-nine years old and occasionally reminding Roger that she is a woman (138), Kathy consistently appears as childish and hasty, and as a foolish girl. She gives in to every desire (her own, and that of strangers), feels disorientated and lost as a tourist, and lacks all forms of security and prospects. Kathy thinks to herself, "you make a fool of yourself you always

---

5 presenting a draft of this chapter in October 2019, I was in dialogue with Julian Murphet. During this conversation, Murphet pointed out that Kathy does not actually return to America in the end of the story. Similarly, Shannon Rose Riley contends that the novel "works structurally to leave the reader at the voodoo doctor [and] it is unclear whether Kathy will ever return to the US" (43–4). I disagree with this reading and suggest instead that the story orientates both *Haiti's* reader and Kathy towards her return to America. I argue that by abandoning Roger and the boys at the pool, Kathy leaves Haiti even if only metaphorically – she cuts her ties to Haiti and is finally ready to move on. Furthermore, the story ends with Kathy's altered state – she is "more dazed than before" and faces the sun (170). Dazed as if she has just experienced a dangerous journey, Kathy survives her quest, and overcomes physical and spiritual darkness. Kathy stares at the sun – the grandiose celestial body and source of life – which acts as a metaphor for America's dominance and capacity to shed light or shadow over its dependent states (like Haiti). As a personification of Acker, Kathy also returns to America in Acker's later texts (e.g. in *Great Expectations*). Therefore, Kathy orientates her journey to America, leaving readers with the impression that she will make it back home despite that *Kathy Goes to Haiti* ends before she manages to do so.

make a fool of yourself everyone's always laughing at you everywhere you go you don't belong anywhere nowhere" (86).

In a different reading, Kathy's naiveté and precarious position are only feigned, and the character is "not as simple as she initially seems." On the contrary, Kathy's promiscuity implies "a new type of gender freedom," and marks Acker's social critique whenever freedom is refused to female characters in the story (Ross, "Review: Kathy"). Or as Chris Kraus surmises, Acker's vulnerability, in both "her writing and life ... was highly strategic" (176). Both images of Kathy highlight some sort of heroic impetus. As naive and vulnerable, Kathy consistently demonstrates her curiosity and capacity to risk. Her recklessness is quite feminist, in fact. Embarking on a journey despite all odds, and in a place where "women in Haiti don't go around alone" (*Haiti* 22), Kathy remains remarkably undisturbed and determined. As the *goodreads* reviewer suggests, she might "get hurt or killed [but] did fine." Indeed, Kathy gets stuck and fucked but this does not trouble her, and for all the hurdles she encounters, Kathy makes it through the end of her holiday unscathed. Undoubtedly, the capacity to survive and proceed hinge on Kathy's luck in the story, and on Acker's decision to keep the plot empty of action and conflict. Yet, a certain dosage of courage underlays all girls' movements in unfamiliar waters, and in sexually-charged spaces:

> I have not ceased being fearful, but I have ceased to let fear control me. I've accepted fear as part of life – specially the fear of change, the fear of the unknown; and I have gone ahead despite the pounding in my heart that says: turn back, turn back, you'll die if you venture too far. (Jong)[6]

The heroism suggested by the second reading lies in women's courage to critique patriarchal oppressions and their power to formulate this critique openly. Kathy continually disproves of misogyny and patriarchy in the story, and this pronunciation has been generally understood as the main function of the character (see Riley, Lee). In fact, Kathy spells out Acker's concerns about female sexuality and emancipation, and dares to do so in pornographic bluntness. Talking like this in the 1970s was certainly an avant-gardist provocation and required courage. We should not forget that it was sexually explicit language that got Acker's *Blood and Guts in Highschool* banned in West Germany and South Africa, and shocked American readers.

---

6 The words are attributed to Erica Jong (the same Erica Jong whom Acker impersonates with her 1982 *Hello, I am Erica Jong*). Despite extensive research, I could not find confirmation that these are actually Jong's words. They appear as an inspirational quote on multiple *Women Solo Travellers* websites (e.g. thehostelgirl.com; teacaketravels.com)

While graphic content should always already be assumed in *Kathy Goes to Haiti* (it is a porn novel after all), the book continues to be recommended as enjoyable iff audiences "set aside any prudish sentiments, step outside the graphic sex scenes, and read with an open mind" ("Review: Kathy"). Protesting in socially inacceptable ways has always proven courage, and Kathy/Acker can be easily described as heroic from this angle. Once we consider Kathy's story and the graphic way Acker tells it, the character's naiveté indeed begins to seem feigned and her frivolity undersigned with feminist disobedience, resolution and strength.

Within the framework of parody, the story is expected to ironize and undermine these images of Kathy. The following analysis shows, however, that the text furthers a vision of Kathy as transformative and forever moving forward despite the mockery and prodigality of foul language and naughty sex acts. The text highlights Kathy's heroic struggles towards "a new world. A new kind of woman. Or a new world for women" (*Haiti* 77). In this sense, the narrative repeats white feminism's ambition to abandon the burdens of the past, visualizes a future that white women like Kathy/Acker can escape to, and identifies the dangers removing this future from the current state of affairs.

> against Black abjection Humans could know *themselves* as agents of change as well as agents who *could* change
>
> (Wilderson, *Afropessimism* 314)

I follow Wilderson here and argue that Kathy's movement serves to bolster the happy ending – Kathy/Acker's renewal reached in the last pages, and with reception of, the novel. In fact, all Kathy's movements, even the most purposeless and parodied ones, are geared towards her survival as a transformative character. Whether she proves to renew herself despite her immaturity, or thanks to her sexual autonomy, Kathy embodies the capacity to transform courageously and stands in direct opposition to what and whom the narrative defines as a burden, as backwardness which paralyses Kathy's desires for "a new world."

It is not too far-fetched to assume that the few white women Kathy meets have shared her enthusiasm and liveliness when they first moved to Haiti. Betty, Roger's wife, complains to Kathy that nothing moves in Haiti, and even dogs or books cannot be imported (61). Although lethargic and stunned when Kathy meets her, Betty remembers and longs for a different life. She exclaims: "[w]hat I really want is a horse. I've always adored horses … If I had a horse, I could go anywhere around here" (60). Yet, living in Haiti (and with Roger) has smothered Betty's energy. She can only talk about her past in America, and fails any opportunity to change her situation in Haiti. Betty is also portrayed as bloodless and imprisoned (61, 66).

In contrast, Kathy yearns to move, improve and change. She feels the urge to emancipate Betty. Kathy says to Roger:

> She's very scared, Roger. She's so scared she almost can't function any more. She doesn't have any blood in her. No one has to be as pale as she is. It's unhealthy. You've got to let her get a job or at least have transportation so she can get out of here now and then. (73)

A moment earlier and a moment later, Kathy sleeps with Roger and asks him to leave Betty. Yet, Kathy's naiveté empties her request of malice and egoism. On the contrary, Kathy consistently behaves in a friendly way towards Betty, and treats her with kindness and understanding. The simplicity and flatness of Kathy's character, and her childish honesty, predetermine her words as straightforward and true. Thus, Kathy appears concerned with both Betty and Roger, and not so inadvertently, with women's liberation from oppressive marriages, the right to pursue happiness, and a new understanding of sex as free, uninhibited pleasure. Kathy's demands for better treatment of women appear courageous, almost self-sacrificial if one considers her naiveté (*here goes the small punk and befriends Betty and has her interest at heart and her husband inside her and it is all a childish play or a feminist protest or a way to steal for the greater good*). Unlike men's fooling around and women's inefficiency, Kathy's movements are marked as intrinsically political. Kathy tells us that she does not want to be a mere tourist whose transitory movements leave the place unchanged (24). On the contrary, Kathy wants to set Betty free, to share with her the benefits of (self)love. In fact, Kathy consistently offers her body to people in need. It is despite her desire, for example, that Kathy sleeps with the taxi driver when she first arrives in Haiti; the man needs physical satisfaction and Kathy presents her body and sex as a gift (11). Kathy thus appears as generous and compassionate, signifying white woman's goodness and generosity. Throughout the story Kathy gives in to everyone, and away everything – her time, money, and body. The result is an endless repetition, complete disorientation. Every day seems the same, every sex encounter repeats the previous. Kathy is stuck:

> Gotta run. Gotta get out. Gotta get moving. Get out. Escape. Escape. Burst open. Stop. Get the fuck out of here anyway I can. Dig my away out ... I want to go home, mommy. I thought this was a passion, but it's not. Emotions are like thoughts. They come and go. They're not me. I can play at being in one, being one, but it's not me, it's just playing, and after a while it makes me sick. I don't know what to do anymore, mommy. Mommy mommy mommy mommy mommy mommy mommy mommy mommy mommy mommy mommy mommy mommy mommy mommy mommy mommy mommy. (87)

To overcome herself, Kathy needs to learn not only how to protect herself and navigate Haitian space, but also how to use what she has to her advantage. It is logical then that narrative resolution occurs when she is told "you have a great force in you. You must go upwards" (165–166). The story thus nears culmination when Kathy realizes she has a strength driving her, that the progress to a better place is intrinsic to her body.

The realization is further enforced by a distinction Kathy draws between her restless self and the sterile, unproductive people she meets in Haiti. Kathy remarks, for instance, that "the land is very flat and women and houses are almost invisible" (110). In fact, the more Acker describes Black female characters as hidden and part of the background, the more feminist appears Kathy's hunger for movement. She is portrayed as capable of change (her mind and destination), and as the only one ready to risk (herself) in the name of bettering the monotonous present. The consistency of this desire throughout the story suggests that moving forward is intrinsic to Kathy's character. Kathy reminds herself: "[y]ou've got to use your intellect to keep you in line ... you're going to go too far out ... you've got to have more cause" (77).

Contrasting Kathy's energy, Haiti is represented by immobile and sick Black bodies and is abstracted through their anonymity and interchangeability. Haitian women appear to be both everywhere (23–24), and nowhere. This contradiction extends Black women's invisibility and makes them part of the background – recognisable only when Kathy feels like noticing them. Like Kathy, Black female characters are marked as girls but, contrary to her, they never grow to be women. Instead, Black female characters appear in the plural, selling sex or cleaning, but never in control of their bodies.

> Black people form a mass of indistinguishable flesh in the collective unconscious, not a social formation of interests, agendas, or ideas
>
> (Wilderson, *Afropessimism* 162)

As Hortense Spillers writes in "Interstices," whiteness re- constructs Black women as "the beached whales of the sexual universe, unvoiced, misseen, not doing, awaiting their verb" (153). In *Kathy Goes to Haiti*, Black women are invisible and useless: neither Haitian men, nor Kathy need them, and the plot continues despite their absence. When Black women appear in the story, they are useless and interchangeable. Thus, even Black women's only characteristic – the capacity to offer sex and satisfaction – is fundamentally different from Kathy's gift of sex and sexual politics. Roger confesses:

> Every day I'm with four or five different women. I spend my lunch break with one woman: maybe I take her to the beach. Then I bring her home in the afternoon. I make a dinner date with another woman and I take her around to the hotels so I can check out what other women are around. Meanwhile I've made a date to meet another woman around ten o'clock in a bar. I go to the bar and spend an hour there. Like tonight I had made a date to meet a girlfriend of mine at Le Poison ... I have dates with women all the time. Sometimes I bring one girl with me to meet another. (148)

Towards Kathy, Roger becomes jealous and obsessive (79–84). He falls in love with her, and makes her promise that she is going to be with him "all the time" (82), and that they "will do everything together" (84). The contrast between singular love and a monotonous overindulgence with available women, marks Kathy as different and better. Sex with her offers something more; she manages to grab Roger's attention and keep it, proving to be superior to the piles of indistinguishable Black women. Furthermore, Kathy reconfigures her subjectivity through the politics and power of sexuality.

"White sexuality is always weaponized"

(Wilderson, *Afropessimism* 195)

Contrasting the monotonous sex of Haitian women, Kathy's sexual encounters are specific and daring. They appear as cutting-edge and an experiment with readers' patience and puritanism, a symbol of white women's new, liberating capacity to reclaim sexuality:

> You've got to get love. You've lost your sense of propriety. Your social so-called graces. You're running around a cunt without a head. You could fuck anybody anytime any place you don't give a damn ... you have this idea lingering from the past maybe you shouldn't fuck so much and so openly other people are looking down on you other people are thinking you're shit. People misunderstand why you act the ways you do. (76–7)

Antithetical to Kathy's urges is the unspoken presence of Black women in the story and their insignificance in and to the plot. While Kathy re-maps identity, moves towards culmination, and carries (in her womb) possibility and future, Black women appear barren, or do not appear at all, barred. Thus, the only time Kathy notices that Black women are pregnant, nursing babies, or aging is at the end of the story when Kathy has already recognized her own capacity to grow. Although pregnant, Black women remain as nameless as before, part of a faceless crowd. Rather than signify Haitian agency, this newly-emerged fertility and presence frame Kathy's maturation and spiritual awakening. First lost and unconscious, Kathy is finally shown to herself (169).

Another indication for Kathy's maturation is her decision to leave the Haitian boys she meets by the pool, and thus her own childishness. Marking youth's joy and playfulness while Kathy was is interested in Haiti, at the beginning of the story the boys raise her curiosity and attract her. Kathy finds their company refreshing and an escape from Haitian men. The longer she stays in Haiti, however, the more burdensome the boys become and the more threatening appears their desire to keep Kathy in Haiti. The boys' sexual needs turn explicit and begin to resemble those of Haitian men. Kathy says to one of the them: "You're too young. I'm twenty- nine years old and you're only twelve." The boy reassures her: "In Haiti there is no age. I'm a man like the Mystere" (107). Soon after, Kathy realizes she needs to grow out of her fascination and fetishism, and to move forward and away. Consequently, she abandons herself as a tourist (and by extension, white female tourism) and chooses to return to the USA – the place construed as harbouring the possibility for growth and renewal (see Berkovitch).

Although Kathy meets an abundance of Black men, their representation nullifies Blackness as much as the absence of Black women in the story. The most successful male characters, the Mysteres, are locked in a tradition of nothingness. Whatever money they make is lost midsentence only for the next generation to inherit the pattern of cyclical gain and loss:

> The grandfather of the present M. Mystere made and lost a million dollars. The grandfather's son made a million and lost a million two times and then went crazy. The present M. Mystere started out life with this heritage and nothing else. Like his grandfather and father, he made and lost a million. (53)

Black men are visualized as interchangeable, a faceless constant. They are depicted as children, and always already contrast Kathy's desire to grow (up). In the effort to satirise Blackness as dull and stiff, Acker uses repeated, tiresome negations. Thus, the more inactive Blackness appears in the story, the faster Kathy moves towards culmination. She thinks to herself:

> I'm the richest prick in town. Poor people don't need arms and legs. I'm the richest prick in town. I wanna go out farther. I wanna get more fucked up. I want to go out there right now. Me go way me me. Moving as fast as the car winds. Grab the cock and up the energy. (98)

Similarly, and in accord with white ideology, Kathy repeats that nothing progresses in Haiti (19, 160). On the one hand, this formal repetition empties meaning; on the other, it piles up meaninglessness itself. A feeling of repetitive nothingness arises in the text – all of Haiti seems like a background and a burden to both Kathy and Haiti's readers. In this light, the form props the story: Kathy struggles to move in a dead space. In this visualization, Black men are part of the scenery and Black women are nowhere.

Kathy's relationship with Roger (she is crazy about him), redoubles the juxtaposition between Kathy's drive and Black incapacity. Her turbulent sexual desires give her an incentive to travel, change, and grow. Conversely, Roger's fascination with Kathy stuns him – he cannot leave his wife, nor offer any prospect to Kathy. Instead, Roger behaves like the boys by the pool and stalls her, refusing to take advantage of the "better" life Kathy promises him. As for her, a brighter future, one without Roger, awaits:

> Someday there'll have to be a new world. A new kind of woman. Or a new world for women because the world we perceive, what we perceive, causes our characteristics. In that future time a woman will be a strong warrior: free, stern, proud, able to control her own destiny, able to kick anyone in the guts, able to punch out any goddamn son-of-a- bitch who tells her he loves her. (77)

It is important to note that Acker reinforces these narrative choices through racializing language as well. She not only tells the story of symbolically nullified Blackness but uses formal experiments to solidify this notion. Thus, the portrayals of Black bodies as naturally incapacitated are relayed in the text as tautologies and repetitions. For instance, Kathy sees "the town [as] absolutely still. A dead mass of houses and shacks and slums piled together, jumbled, shackled, no reason at all, just there" and reiterates that this can be nothing else but "death. Stillness" (98). To settle this formally, Acker employs a stubborn repetition:

> *men and women* and *girls and boys* and *babies* sit and argue and sell and buy and *stand around* and eat and *walk* ... the shacks begin again and *all the people walk and sit* and talk and carry baskets and have *dogs* and quarrel ... There is *dust* everywhere. *Dust* on the road, *dust* in the air, *dust* on the skin, *dust* on the straw and wood shacks. The distance is a *light tan* haze. The shacks are *light* tan and grey ... *Women and a few children walk* in these ruts ... *dogs* and chickens run from the shacks *into the street* (...) the *road's* a hard *dirt road.* It *winds around, goes up and down,* basically it *moves north-south...*The *road's* flat and runs directly *north and south.* (9, italics mine)

The repetition appears in the description of Black space and is a good example of how anti- Blackness does not get broken by the postmodernity of Acker's aesthetics. It is a repetition that exhibits the shared imaginations and stereotypes that Guinn praises Acker for dismantling. Acker displays Blackness as homogenous, leading nowhere, and invisible – a display Guinn's argument relies on, and frames as a parody that succeeds in its political aspirations. Contrary to Guinn, I argue that this piling of words adds to dehumanizing Blackness by accumulating its descriptions and rendering them meaningless. The meaninglessness echoes Acker's portrayal of Haiti because the textual space dedicated to the landscape seems as contagiously repetitious as Black characters. Put otherwise, the text reiterates the emptiness Kathy sees in Haiti.

Acker's textual lingering achieves a feeling of stasis and disorientation. At the same time, Kathy's descriptions parallel her restlessness: at one time she is joyful and without a care in the world, then she jumps into breathless monologues. Kathy reproaches herself:

> You step on the people you meet. You use them for your insane desires. You don't know the difference between friends and strangers and you're unable to give anyone, especially the people who say they want to fuck you, ordinary human affection. You're beyond the bounds of being human. You're inhuman. You are now out of control. (77)

For critics like Guinn, Acker's narrative choices are successful in their intentions. Like Guinn, Victoria de Zwaan sees Acker's parody as a recycling of the original text "into a more deliberately and overtly intertextual metafiction" ("Rethinking"). De Zwaan follows Istvan Csicsery-Ronay Jr in asserting that Acker's recourses to parody should not be restricted by narrow definitions of the technique. While I generally agree with this, I want to stress that textual repetitions in *Kathy Goes to Haiti* perform differently than those in Acker's other appropriations. One reason for this is that *Kathy Goes to Haiti* offers no alternative representation to a stereotype used to justify colonial violence. As Steve Martinot and Jared Sexton argue, the "repetition of derogation becomes the performance of White supremacist identity" (5). In a similar vein, Denise Riley notes

> In its violently emotional materiality, the word is indeed made flesh and dwells amongst us – often long outstaying its welcome ... On occasion the impact of violent speech may even be recuperable through its own incantation; the repetition of abusive language may be occasionally saved through the irony of iteration, which may drain the venom out of the original insult ... Yet angry interpellation's very failure to always work as intended ... is also exactly what, at other times, works for it. (9–10)

In *Kathy Goes to Haiti*, the exaggeration of racism, which is Acker's intended parody, fails the moment it re-produces the very image Acker set out to criticize. Objects and subjects in the story are inconceivable outside Acker's visualization. Thus, the story forces Black characters into continuous symbolic colonization through its re-readings. As Saidiya Hartman argues in "Venus in Two Acts," the violence exerted upon Black bodies is suspended in repetition insofar as their narrated dis/appearance redoubles the scene of oppression and becomes the only way they are summoned by imagination. She further asserts that the stories that exist are not about them, but rather about the violence, excess, mendacity, and reason that seized hold of their lives, transformed them into commodities and corpses, and identified them with names tossed-off as insults and crass jokes. (2)

Admittedly, *Kathy Goes to Haiti* was never intended as a realistic account. It is only an artistic intervention; it parodies reality in Acker's original way. However, the particular focus on, and representation of, Blackness is not only Acker's. In fact, projects like *Kathy Goes to Haiti* continue to appear and showcase white women's political art (or creative politics). Let us take for example Dana Schutz – a white contemporary artist who produced a piece in response to gun violence in the US. Like Acker, Schutz has publicized her interest in art as the "place where the hierarchies of the world can be rearranged" (artnet). Put otherwise, Schutz also sees her work as an opportunity to expose and re-imagine reality.

In her 2016 painting "Open Casket," Schutz portrays Emmett Till – the Black boy who was lynched by two white men in 1955. Like *Kathy Goes to Haiti*, the painting zooms in on Blackness. In fact, Schutz repeats the movement of Acker's gaze – her painting depicts the intensity of a dead(end) Black body. Schutz's work is not a parody like *Kathy Goes to Haiti* but its image is constructed in a similar way – Blackness is highlighted, it is where the white artist goes to articulate social criticism. In an open letter to the Whitney Biennial, which showcased Schutz' painting in 2017, artist and writer Hannah Black writes:

> I am writing to ask you to remove Dana Schutz's painting "Open Casket" and with the urgent recommendation that the painting be destroyed and not entered into any market or museum … the painting should not be acceptable to anyone who cares or pretends to care about Black people because it is not acceptable for a white person to transmute Black suffering into profit and fun, though the practice has been normalized for a long time … a similarly high-stakes conversation has been going on about the willingness of a largely non-Black media to share images and footage of Black people in torment and distress or even at the moment of death. ("The Painting")

Black writes further that although Schutz might have been motivated by shame of the murderous arithmetic of whiteness, her artistic appropriations and use of Black suffering as "raw material" derive from that very arithmetic. Hannah Black is not alone in deeming Schutz's work problematic. The open letter was accompanied by serious criticisms against the fetishization of Black pain, the ease with which (white) morality is read into anti-Black practices, and the feigned naivete with which white cultural producers continue to utilize Blackness (see Ziyad, Muñoz-Alonso).

> Black misery is profitable for racists
> and
> anti-racists
>
> (James, "Reaching Beyond")

Recalling the work of another white artist – Ti-Rock Moore – Hari Ziyad reminds us that Schutz is only one of many who capitalize on Black suffering. "Under the hand of the likes of Schutz and Moore," Ziyad writes, "Black death becomes less a call to awareness, more a titillating spectacle" ("Why"). Again:

> [Black people] *are* being genocided, but genocided *and* regenerated, because the *spectacle* of Black death is essential to the mental health of the world – [Black people] can't be wiped out completely, because [Black people's] deaths must be repeated, *visually*
>
> (Wilderson, *Afropessimism* 225)

At the heart of the question lies not censorship as Coco Fusco suggests ("Censorship"), but an unredeemable distance between white and Black positionality (Wilderson, Fanon). This distance cannot be bridged because

> the embodiments of opposing and irreconcilable principles or forces ... hold out no hope for dialectical synthesis, and because they are relations that form the foundation on which all subsequent conflicts in the Western Hemisphere are possible. (Wilderson, *Red, White, and Black* 29).

In other words, while themes and images may share similar contours (Fusco excuses Schutz' work with the fact that both white and non-white artists have represented Black suffering), the cultural work and imaginative labour embedded in, and seeping from, these images is quintessentially different. Eve Tuck and C. Ree say in one voice:

> Damage narratives are the only stories that get told about me, unless I'm the one that's telling them. People have made their careers on telling stories of damage about me, about communities like mine. Damage is the only way that monsters and future ghosts are conjured. (647)

There are two meanings of "damage narrative" implied here. Firstly, damage points to the focus of representation, to the dominant visuality that highlights and prefers to see hurt (the hypervisualization of Black suffering to which Hannah Black refers). In Schutz's example, this is Emmet Till's death; in Acker's – Haitian poverty and social paralysis. Hortense Spillers exposes this focus and mode of representation as the colonial practice of reducing Black bodies to flesh for white voyeuristic pleasure and spectacle (what Spillers calls in "pornotroping" in "Mama's

Baby"). Schutz's painting and Acker's *Kathy Goes to Haiti* both cash in on damage; they are damage narratives because they derive "profit and fun" forcing pain to the spotlight.

The second meaning of damage is contained in fact that the story is not told by the person or community experiencing suffering: "damage narratives are the only stories ... unless I am the one that's telling them." In this sense, a damage narrative is one that produces further damage, it is an oppressive exertion of white imagination onto non-white bodies. The violence of reproducing an already real scene of violence is doubled in white authorship/authority to claim Black bodies ad infinitum (Aldridge, "Black Bodies White Cubes"). From this perspective, Acker's saturation of stereotypes about Haiti can never possibly un- or re-do the "image" itself (the reference cannot go away by being ironized). On the contrary, the act of anti-Blackness is locked before, and in, the text. Postmodern form and parody might, indeed, abandon realism, but there is a reality here (or, rather a white construction of reality) that is in excess of Acker's technical ambition. The reception of *Kathy Goes to Haiti* also points to this excess. Literary criticism has praised Acker's creativity and fight for freedom. As Clare maintains,

> Acker's insistence that freedom is a process and not a product contributes to a body of work that always tries to imagine something new or different ... and keeps open the possibilities for change. ("Why Kathy")

Calvin Warren suggests, however, that it is often "black suffering and death [that become] the premiere vehicles of political perfection and social maturation" ("Black Nihilism" 7). From this perspective, the way *Kathy Goes to Haiti* ironizes notions of Blackness as lacking the capacity to generate anything only redoubles white mythology – both in redoubling racist stereotypes, and of the celebration of women like Acker. Not surprisingly, the rare considerations of race in *Kathy Goes to Haiti* have focused on Acker's ethical and political triumphs. Shannon Rose Riley claims, for instance, that Acker "sticks a little knife up the annals of the U.S. national imaginary" and *Kathy Goes to Haiti* is

> a nagging reminder of the almost entirely erased history of the U.S. military occupation of and of long-standing white anxieties in the U.S. about interracial sex and desire. (34)

Marilyn Manners commends Acker for her "alternate episteme" and the "dismantling of the universality of whiteness" (108). Riley and Manners do not address the problems emerging with the teleology of white renewal. They also miss the fact that its narrativization is often parasitic in relation to Blackness and built through an antagonism best abbreviated Transformative Whiteness – Static/Backward Blackness (Morrison, Spillers). As Sherene Razack asserts, "[w]e know the black body by its immobility and the white body by its mobility" where mobility is always already orientated towards, and indicative of, (new) being (13).

The point I want to push is that the radical sexuality Acker narrates – Kathy's capacity to transcend patriarchal claustrophobia and her initial reason for moving – is realised on Black flesh and through Black fetishism. This is not a minor transgression. Where Kathy arrives as a new woman is where Acker redeems white subjectivity and where she fixes Black being. Thus, whatever the intention behind parodying Blackness as lifeless, its negations remain the only way it summons representation. We, as readers, can only acknowledge the lack, the misnaming, but this is where we are left off, too. Here Black bodies are made doubly fungible – once, to sustain the parody and a second time in Acker's own writerly success. As Hartman explains in *Scenes of Subjection*

> the fungibility of the commodity makes the captive body an abstract and empty vessel vulnerable to the projection of others' feelings, ideas, desires, and values; and, as property, the dispossessed body of the enslaved is the surrogate for the master's body since it guarantees his disembodied universality and acts as the sign of his power and dominion. (21)

Kathy Goes to Haiti situates Kathy in the middle of anti-Blackness: Kathy travels to Haiti to immerse herself in Blackness, exploit Black bodies, and by contrasting their "inferiority" and "backwardness," shines through (as a new woman) and re-imagines herself. The premises of Haitian Blackness remain intact. Nothing happens in Haiti after Kathy leaves and even Roger, whom Kathy has desired so passionately, disappears from the narrative unnoticeably. Kathy remains the same but better; she is "more dazed than before" but happy, calm, and in control of her impulses (170). Haiti/*Haiti* highlights her new self. In fact, Acker's text extends the "structure of banality" as Martinot and Sexton say, and it is this structure of banality against which Kathy's transformation and movement stand out as singular and spectacular, as truly new. Kathy might be a broke naïve girl but journeys all the way to Haiti, and against all perils manages to turn a deadly space into a happy daze in the sun.

Exit routes

Presenting my initial thoughts on this chapter at conferences, and later as an article submission, I faced several important criticisms. These contentions are important to mention here because they reveal why a disavowal of Acker and her work, which is how my contribution has been received by some, has displeased my critics' desire for a healthy hero, one that successfully abandons the burdens of the past and of their own limitation. If the offered analysis has been convincing to my readers, however, these criticisms point also to which transgressions are deemed ignorable and harmless for Acker's image as a good (in the meaning of well- intended and successful) emancipatory writer. I would like to address the disagreements partly to continue the conversation, and as a way of opening this chapter's conclusion to a more general meditation on how words, images, and ideas move among us.

Most importantly, my arguments were criticized for conflating Acker and the main character in *Kathy Goes to Haiti*, and respectively, of their (textual and discursive) renewal. This conflation was deemed unsavoury. Contrary to this view, I emphasize the facts that 1) Acker continuously used language to de/construct her identity, and overlaid biography and fiction,[7] and 2) in *Kathy Goes to Haiti*, she abandoned experimentation, and wrote it as a literal text.[8] Acker also used the 1976 CAPS grant[9] – the only grant she ever received – to travel to Haiti and do ethnographic research for the novel. Acker named the main character after herself, and modelled Kathy's movements on her own. I agree, of course, that narrative representation always already exists at a distance from what is narrated – especially for a devious and disobedient writer like Acker. As Michel Foucault notes, representation is "perpendicular to itself: it is at the same time indication and appearance, a relation to an object and a manifestation of itself" (72). As quickly as Acker disowns the novel and what it represents, however, as readily she admits: "when I write, I don't write about things, I do them ... that's visceral" ("Interview").

Criticism against my aligning of Kathy and Acker arises with reading the first as a parody, something the second – Acker – invented, in reverse proximity to herself, and for the purposes of irony and deconstruction. As a parody, Kathy's character is taken as a function of Acker, and the character's transgressions as fictional and limited to the story. As my analysis shows, however, in a certain light parody can be seen to overflow narrative intention, spill away from its purpose and execution, and reach the limits of deconstruction. It is important to note that I read Kathy and Acker together less out of caprice and sarcasm, and not because I confuse narrator and author.

---

7 Note for instance that even biographical milestones such as Acker's birth year are contested due to Acker's play/provocations with these (Kraus, Colby). Kraus underscores just how intensely Acker experimented in reality and writing to create a 'narrative myth' she could live and embody. Thus, 'Acker's life was a fable' but its 'facts are hard to pin down in any literal way. Because ... Acker lied all the time' (14)

8 Acker included *Kathy Goes to Haiti* in the series 'Literal Madness'

9 The Creatives Artists' Public Services Program New York State Grant for Fiction

Rather, conflating in some measure Kathy and Acker allows me to highlight how dominant readings overlap and re-adjust the lens. This is evident in Acker's reception but also in that of Don DeLillo as elaborated previously. Much can be said about the particular works and authors – in the very least, that they thematize a desire for renewed consciousness, and do so in their own avant-gardist spirit. More crucial is the question why texts which organize renewal through repertoires of anti-Blackness – something the novels and their narrators hardly conceal – have failed to problematize the transformative potential read into writers like Acker and DeLillo, and to raise doubt in their antiracist engagements. On the contrary, the authors' reputations have navigated interpretations of the stories and somehow absolved their characters' mis/conduct.

In Acker's case, an idealization of her formal and ethical achievements goes against her frequent explorations of failure, and the fact that failing "is something queers do and have always done exceptionally well" (Judith Halberstam, "The Queer Art" 3). Critical of my contention that both Kathy, the narrator, and Acker, the author, use Blackness to centre white women's renewal is criticism that allows the transgression of the first, and disallows it for the second. Obviously, this disconnect interests me deeply. Are Acker's oeuvre and personae compelled to perform as effective, and at what cost? How serious do avant-garde circles treat an injury on Blackness, and can a play of words, stereotypes, and ideological markers really "do nothing"?

Indeed, white criticism has embraced Acker as a controversial, and even self-contradictory figure (see Glueck, Kraus, Borowska). She continues to be remembered, however, as a subversive hero, one who courageously defies oppressive ideologies and embodies transformative potential. My analysis shows that this memorialization transfers onto *Kathy Goes to Haiti* despite its indebtedness to white ideology, and the failure to ironize and deconstruct it. Moreover, Acker's image as an insurgent writer regulates interpretations of the text – it pushes forward the text's alleged anti-racist program, and conceals (in plain sight and blatant parody) the much more complicated and implicated position Acker inhabited in her texts and life.

Black critique has long insisted on re-considering positions like Kathy/Acker's, because dominant language is amongst the practices of violence that render racism permissible (to the body and to the imagination). Exposing the intersection between colonial violence and literature, Toni Morrison theorizes the necessity "to free up the language" (*Playing* xi). As M. NourbeSe Philip shows, language itself is always already "contaminated, possibly irrevocably and fatally" (199). How then does Acker's iteration of colonial stereotypes sit with her rebel politics and love for outcast bodies? Furthermore, what do practices of reading mean in their multiplications of past images and narratives? Like DeLillo's narrator,

Kathy asks of her readers to imagine in the same old light: this is the premise and finale of her parody.

Unlike DeLillo's narrator, Kathy goes straight to the bodies who excite her. Acker's novel was published nearly forty years before DeLillo's, in a different time and for a different audience. Admittedly, the saturation of Blackness in *Kathy Goes to Haiti* differs from DeLillo's exclusion of Black characters; the difference is blunt and striking at the same time. Yet, both novels centre white heroes and their movement toward a new world by ricocheting off Blackness. In both *Zero K* and *Kathy Goes to Haiti* Blackness functions as a fixed reference point – something the protagonists overcome, and against which they prove transformative. Blackness thus marks the heroes' capacity to transcend, to be/create new(ness). Acker and DeLillo's narrations use Blackness as a literary invention, as an instrument to the plot. Perhaps because of this, and the fact that *Kathy Goes to Haiti* and *Zero K* are often taken for minor and niche, the present engagement with both novels has been criticized as fortuitous and idiosyncratic.

Yet, there is a narrative that runs parallel to Acker and DeLillo's. It is more mainstream than either of the novels I discussed. When the 45th President of the United States Donald Trump calls Haiti a "shithole country" – stricken by dysentery Kathy communicates the same sentiment– and suggests that Haitians should be taken out because America does not need them – DeLillo's exclusion of Black bodies communicates the same sentiment, too – Trump repeats the basic premises of *Kathy Goes to Haiti* and *Zero K* with a seriousness often severed from fictious texts. The trope of the shithole country and the practice of eliminating Blackness have proven attestable in pronunciations like Trump's because he is, in personifying whiteness, the "first white president," the recognizable enemy of left politics (Coates "The First").

With this section, I have insisted on revisiting Acker and DeLillo's texts not only because their usage of language parallels what Samira Saramo would call a Trumpian rhetoric,[10] i.e. white ideology, but also because there is something to be said about the practices which safeguard leftist figures and distance them from complicity in anti-Blackness. From this perspective, we can see that images move, often only to align with available repertoires, and always already regulated by dominant visuality.

---

10 See also Saramo's "The Meta-Violence of Trumpism."

3 Against dominant visuality, or the cut as critique*

Cutting the daguerreotypes of two Black people in her book *In the Wake: On Blackness and Being*, Christina Sharpe performs the subversion of *worrying the line* as Cheryl Wall translated once the blues trope (*Worrying*). The daguerreotypes Sharpe destroys are part of an ethnographic series ordered by Louis Agassiz in the 1850s. Portraying enslaved men and women, the series aimed to "reveal what blackness looks like and how to look at blackness" (*In the Wake* 43). The images Sharpe revises, or "redacts" as she terms it, are those of Delia and Drana – two enslaved people born in America. Sharpe explains:

> In a move that is counter to the way photographic redaction usually works – where the eyes are covered and the rest of the face remains visible – here I include only Delia's and Drana's eyes … I redact the images to focus their individual and collective looks out and past the white people who claimed power over them and the instrument by which they are being further subjected in ways they could have never imagined or anticipated. I want to see their looks out and past and across time. Delia and Drana. In my look at them, I register in their eyes an "I" and a "we" that is and are holding something in, holding on, and held, still. Delia and Drana sitting there (still) and then standing there (still), and clothed and unclothed (still) and protected only by eyelashes (still). (118)

The cut "annotates" rather than erases the images. Sharpe takes a violent portrait, subverts it, and shows us more by showing less. Although the cut adds no concrete knowledge – for this is a petrifying impossibility (Hartman "Venus") – it reveals that something other is there, beyond reach and representation, "in some other place already, less tortuous, less fleshy" (Brand, *In Another Place* 247). Altering the image, Sharpe denies Agassiz's authority to visualize Black bodies and the lines of the visualization. She destabilizes the image, shifts accent and focus, but also highlights the act of seeing itself. Thus, the old portrayal is refused, and in its place appears not only a new one but the tension of representation and sight. The viewer becomes aware of their own act of seeing; before our eyes appear not only Delia and Drana's but also the history that Sharpe refuses (*In Beloved, Sethe cuts the throat of her daughter because she cannot bear to see her live as a slave; the cut is refusal and care*).

Sharpe's redaction can be understood as an optical expression of what Tina Campt called "listening to images." Like Sharpe, Campt goes against dominant discourse and insists on seeing otherwise: she revisits old photographs but "questions the grammar of the camera," engages in a "haptic form of sensory contact," and employs "the endlessly generative space of the counterintuitive" (*Listening* 6–7).

https://doi.org/10.1515/9783110799996-004

On the one hand, Sharpe and Campt call into question the specifics and certainty of images, the history that produced them, and the injury and limitations of the archive. On the other, they make noticeable how the retina moves, as if giving it a sound so that each change of direction, each pause, can be registered (Sharpe "Response"). I understand the revisions as critique not only of anti-Black representations but also of anti-Black acts of seeing: the way one lines up ideas and pixels until they frame one coherent shot. Martinot and Sexton explain that

> between the inability to see [past anti-Black vision] and the refusal to acknowledge [frequencies beyond anti-Blackness] a mode of social organization is being cultivated for which the paradigm of policing is the cutting edge. (172)

Policing kills (Wilderson "We're Trying") which is why Sharpe redacts the image with refusal and care, in the wake and as wake work:

> living in the wake means living in and with terror in that in much of what passes for public discourse about terror we, Black people, become the carriers of terror, terror's embodiment, and not the primary objects of terror's multiple enactments; the ground of terror's possibility globally. This is everywhere clear as we think about those Black people in the United States who can "weaponize sidewalks" (Trayvon Martin) and shoot themselves while handcuffed (Victor White III, Chavis Carter, Jesus Huerta, and more), those Black people transmigrating the African continent toward the Mediterranean and then to Europe who are imagined as insects, swarms, vectors of disease; familiar narratives of danger and disaster that attach to our always already weaponized Black bodies (the weapon is blackness). (15–6)

This is also clear in the texts I discussed earlier. DeLillo and Acker's accounts mark Blackness as abstract and all-encompassing danger. Like Acker's Kathy, DeLillo's girl – the embodiment of human failure and fragility – rides her bike not expecting the disaster on the horizon, the *Thing* which will swallow her whole. Using textual negations (Black bodies are erased, misrecognized, and piled until they lose contours and content) and framing Blackness as negative (Black bodies are described as inferior and unhuman), Acker and DeLillo develop a portrait with available fictions and techniques. They extend, in often-praised efforts to escape, anti-Black representations. The latter prove durable in opposite registers: absence and presence both carry meanings pulled as if by gravitational force (*Sharpe says anti-Blackness is the totality of our environment, it moves everything as invisible, invincible force [104]*). Flattened by the modalities of "there" and "not there," Black characters are thrust into symbolic colonization. To be included or not rests on the same violence because the modalities (there/ not there) are as constricting and insufficient just as the system itself. Which is why *a cut* can work as an opening (Philip 189–207) and here is how Jack Halberstam reads Harney and Moten's *The Undercommons:*

> If you want to know what the undercommons wants, what Moten and Harney want, what black people, indigenous peoples, queers and poor people want, what we (the "we" who co-habit in the space of the undercommons) want, it is this – we cannot be satisfied with the recognition and acknowledgement generated by the very system that denies a) that anything was ever broken and b) that we deserved to be the broken part; so we refuse to ask for recognition and instead we want to take apart, dismantle, tear down the structure that, right now, limits our ability to find each other, to see beyond it and to access the places that we know lie outside its walls. We cannot say what new structures will replace the ones we live with yet, because once we have torn shit down, we will inevitably see more and see differently and feel a new sense of wanting and being and becoming. What we want after "the break" will be different from what we think we want before the break and both are necessarily different from the desire that issues from being in the break. (6)

By breaking dominant representation, Sharpe's cut opens space beyond it. Beyond is where Delia and Drana are standing still, loved and not alone. Sexton reminds us that "Black life is not lived in the world that the world lives in, but is lived underground, in outer space" ("The Social Life" 28). Outer space, or other space, is where anti-Blackness ceases to regulate:

> My whole body changed into something else. I could see through myself. And I went up ... I wasn't in human form ... I landed on a planet that I identified as Saturn ... they teleported me and I was down on stage with them. They wanted to talk with me. They had one little antenna on each ear. They talked to me. ("Space is the Place", qtd. in Szwed)

With the above words, the jazz musician Sun Ra imagines a space wherein he is no longer confined to the world as we know it. He imagines outer space in "freedom from the world, freedom from Humanity, freedom from everyone" (Wilderson, *Red, White, and Black* 23). For Sun Ra, his body, not only his language, is codes-witching – his body means something else Black critique reconstructs what it means to be Human (Weheliye, Wynter, Spillers). The stage is "down" (or, beyond/underground) and Sun Ra is not seen (as spectacle) but heard. He has also performed a cut – sometime in the 1950s Sun Ra has redacted his slave name and instead chosen Le Sony'r Ra. With the cut comes jazz – Black music which refuses anti-Blackness and the world (Okiji, Moten) and to which Sun Ra dedicates his life. Of his work, Klaus Thiel writes:

> Ra's texts are different. What is a text? Something that was made for repetition. What is repeated? sentences? words? letters? syntagmata? graphemes? What Ra does, among lots of other operations, is to cut. Every way of cutting. (24)

Sharpe tells us that the practice of the cut, the cut as critique, is unflinching and personal ("an "I" and a "we" that is and are holding something in, holding on, and held, still"). To cut is different than to erase. Of Kara Walker's cut-outs, Sharpe says:

> I read Walker's cut-outs as representing a violent past that is not yet past in such seductive forms that black and white viewers alike find themselves, as if against their will, looking and looking again. (*Monstrous* 156)

The mother of Emmett Louis Till cuts protocol. It is her son's funeral, and Mamie Till Bradley demands her son's body to be returned to Chicago and be buried in an open casket *look and look again*. Unlike Schutz' "Open Casket," Mamie Till challenges dominant visuality and violence with care and wake work:

> Mamie's statement was so powerful because Emmett was her baby. She chose Black media specifically to reprint the images. But under the hand of the likes of Schutz and Moore, Black death becomes less a call to awareness, more a titillating spectacle, using non-Black media and galleries to recreate images of it with no regard to the fact that they were once lives–once loved. (Ziyad, "Why")

According to Ziyad, Emmet Till's mother subverts the coherence of the world in which her son appears as threat and operates on a level different than visualization.[1] Ismail Muhammad writes

> These same images ... initiate us into a black community characterized by creative acts of defiance – such as Mamie Till's gesture – that create "counter-memories," or correctives to traumatic histories. ("On Seeing")

Cutting protocol, reframing and refusing, Black communities break dominant discourse into and through "acts of defiance." Reading Black criticism, one understands that defiance occurs at "a lower frequency" (Gilroy 37), and as a spectre/spectrum of refusal that whiteness cannot claim or understand. Which is why Black critics like Sharpe and Moten theorize the cut as an alternative space that exists despite and against the white world. "There is an ethics of the cut," Moten writes, that "instantiates and articulates another way of living in the world." Moten asserts further that this is

> a black way of living together in the other world ... in the alternative planetarity that the intramural, internally differentiated presence—the (sur)real presence—of blackness serially brings online as persistent aeration, the incessant turning over of the ground beneath our

---

1 see #iftheygunnedmedown

feet that is the indispensable preparation for the radical overturning of the ground that we are under. ("Blackness and Nothingness" 778–9)

"Can you hear a shadow?" asks Nicolas Brady and challenges visuality so that it no longer colonizes the entirety of being (there or not there) ("Louder"). Sun Ra imagines being heard, playing in outer space, somewhere beyond. Black play takes shape "outside the reaches of your white understanding" writes Susan Lori Parks in "New Black Math," and here is what Sun Ra writes in "Cosmic Equation":

Then another tomorrow
They never told me of
Came with the abruptness of a fiery dawn
And spoke of Cosmic Equations:

The equations of sight-similarity
The equations of sound-similarity

Subtle Living Equations
Clear only to those
Who wish to be attuned
To the vibrations of the Outer Cosmic Worlds.
Subtle living equations
Of the outer-realms
Dear only to those
Who fervently wish the greater life.

(qtd. in Geerken 110)

Subtle and outside white understanding, wake work continues. *Strong eyes, protected by eyelashes, resist the world:* "I look hard at her so she will know that the clouds are in the way I am sure she saw me I am looking at her see me she empties out her eyes" (*Beloved* 211).

Futurity

4 Ten easy steps to inherit the future. Visualizing renewal and the old prodigal boy in Marylinne Robinson's *Gilead*

Overview

In the following, I consider renewal as a promise and premise of the future – as a *thing* that happens in a future time but also as something that generates futurity. As I have mentioned earlier, dominant representations of futurity do not refer strictly to a time to come, to what will be hereafter, but entail the "prospect of a better tomorrow," a renewed state of being that "may prevent our world from closing in" (Eshel 5). Black and Gender critique show that dominant representations of the future always already exclude marginalized groups (see Campt, Munoz, Keeling). This exclusion becomes visible if we examine 1) how futurity is arrived at in narratives and as an image/imagination, 2) and what happens to alternative conceptualizations of the future and critique of its racially exclusive premises. The capacity to renew underlays both these issues: renewal is directly linked to and shaped by the ideologies navigating the future as a concept, and is crucial for the de/construction of this concept.

My main interest is to see how dominant representations of futurity interplay with alternative ones (i.e. critical of dominant discourse), and how white ideology works to cancel subversive visualizations of the future. I will discuss dominant (white) and alternative (Black) representations that are already in play. This play clarifies how dominant discourse attempts to "eat up" and silence alternative visions, an attempt I call *rewording* for the lack of a better term. I argue that by rewording, dominant discourse incorporates yet defuses critical interventions, i.e. it twists their meaning, changes the lighting, and points the lens elsewhere. Before I analyze this process, I consider the construction of futurity in Marilynne Robinson's highly praised novel *Gilead.* [1]

Written as a letter a dying father leaves to his young son, *Gilead* thematizes the question of renewal/future explicitly. The father, Ames, recounts his life and tries to imagine but also "re- appear" at a distant point in time (the time his son will read the letter, the time the son experiences without his father). I focus on Robinson's

---

[1] The novel received the 2005 Pulitzer and the National Book Critics Circle Award; there also seems to be something like a consensus over Robinson's gracious, soothing language and style (Weele, Shy, Lear); Robinson is also celebrated for her talent to "humanize" tradition (Shy 254)

https://doi.org/10.1515/9783110799996-005

narrativizations of the future, and examine whether and how the future is regenerated/renewed in these narrativizations, and how it relates to and borrows from dominant discourse. *Gilead* is instructive in this regard, for while it illustrates one dominant vision of futurity, it also references a text that has offered an alternative one – Toni Morrison's *Beloved.* This is the play, or rewording, I want to consider – what happens to Morrison's ideas when they are incorporated in Robinson's text; do they change, and if yes, how?

To this end, I first examine Robinson's construction of futurity and then move to analyze her rewording of *Beloved.* As I will show, dominant discourse frames white time as promising (full of prospects for a better tomorrow and progressing towards it) and Black time as static (empty of life and leading nowhere). Through rewording, Robinson erases a different view on future time and the critique of her endorsed and exclusionary conceptualization of futurity. From the vantage point of Afropessimism, *Beloved* is a text which clearly articulates such critique (see Wilderson, Murillo). On the one hand, the story exposes the "always now" of an unlivable reality as produced in/by the white world (Morrison, *Beloved* 210). On the other, it "opens" no/time by bringing back a dead child, Beloved – that strong is Beloved's will to live and love for her Black mother, that strong is her protest against a world that disallows Black futures – so strong that it makes her rise from the dead.

By looking at what Robinson changes in Morrison's text,[2] I consider the interplay between the two representations of futurity. To make my argument clearer, I continue the discussion by shifting my focus towards another setting: the discourse around the "Future" billboard by the Black artist Alisha B. Wormsley and the billboard's forced removal. Once again, my interest revolves around the effects of white work (white ideology working) in dominant discourse's engagement with Black critique. As my analysis shows, white people not only rejected Wormsley's work by taking the billboard down but reworded, i.e. changed its meaning, and thus emptied its political potential. Drawing a connection between Robinson's rewording of *Beloved* and the rewording of Wormsley's message, I point to one of the ways whiteness racializes steady images (like that of the future) and the capacity to produce them, i.e. to produce (new) knowledge and language, and new grammar to order them in.

---

2 I follow John Murillo in a rather "*specific* reading [of *Beloved*] from what is and has been a primarily Afropessimistic framework" (95). I do so departing from the vast and necessarily different interpretations present and possible when one analyses the novel with another focus and a different theoretical repertoire. Like Murillo, I embrace this "alternative reading" in its Afropessimistic momentum for tackling questions regarding "Black time, space, and creation" and their antagonistic relation to white time, space, and renewal.

More broadly, the chapter aims to consider one of the ways whiteness nullifies Black critique against white domination in/of time, and how it obstructs alternatives by simultaneously idealizing the same old impetus to (unknown-but-white-nonetheless) futures. A worthwhile ending of this chapter would have been to examine critical visualizations of futurity – for example in the works of Octavia E. Butler and Nalo Hopkinson. It will also be interesting to consider "The Comet" by W.E.B. Du Bois or the canvases of Jean-Michel Basquiat, or movies like Frances and Nuotama Bodomo's *AFRONAUTS* and projects like *Black Futures Lab* and *Afrofuturist Abolitionists of the Americas*. Such analysis is beyond the scope of this study and, in my opinion, cannot be accomplished by a white reader.

Instead, at the end of *Futurity* I return to *Beloved*. This section references some of the ways Afropessimist readings frame *Beloved* – framing that lies in stark contrast to Robinson's rewording of the novel. In a calibrated * passage, I join critics who see *Beloved* as protest against the deadly futures whiteness prescribes for Black people, and as Black Beloved Power to fight and rise above whiteness's murderous arithmetic. From this perspective, *Beloved* does not spell out an alternative way of being in the future; it does not draw a concrete image of a future free of whiteness, at least not one that we can order within the binds of imagination. Exposing the ongoing violence of whiteness (among which anti-Black claims on futurity), however, the novel works to undo the restrictions and regulations of time. By fighting the existent limitations to liveable Black futures, *Beloved* seems to be directly concerned with futurity – even if the narrative world does not take us, readers, further ahead but instead brings us, and Beloved, back.[3]

Working outside normative ideas of time, Beloved comes back and exposes their dead end. From the perspective of Afropessimism, the destruction of each white mode for existing in and making sense of the world is necessary, it is indeed the only way to open space for a truly new world:

> Nevertheless, the slave is a sentient being. Therefore, an existence void of transformative promise, which narrative holds out to human subjects, is a painful lesson for the slave to learn, much less to accept. I am not suggesting that Black people should resign themselves to the inevitability of social death – it *is* inevitable, in the sense that one is born into social death just as one is born into a gender or a class; but it is also constructed by the violence and imagination of other sentient beings. Thus, like class and gender, which are also *constructs*, not divine designations, social death can be destroyed. But the first step toward the destruction is to assume one's position (*assume, not celebrate or disavow*), and then burn the ship or the plantation, in its past and present incarnations, from the inside out. (Wilderson, *Afropessimism* 103).

---

3 See Murillo, 95–120.

I read Morrison's Beloved as a character who is destroyed and thus embodies destruction (or to follow Murillo, an *avatar of destruction*) but whose return explicates the urgency to get rid of the white world and white world-making – for the sake of beloved Black children who demand life other than the forced death this here world presents them with. "This here new Sethe," writes Morrison of Beloved's mother, "didn't know where the world stopped and she began." (164) That there is a different beginning, an alternative way of being beginning with the end of the white world, is an idea that opposes Robinson's idealizations of ever renewable, sustainable white futures. In the following, I consider these two irreconcilable visions, and show how white futurity always already works to negate Black life.

The infinity mirror

Some outcasts are meant to be saved for a brighter future. Like Acker's Kathy, John ("Jack") Boughton feels dazed after his blessing. He is not in Haiti, but in Gilead – the fictional town of Marylinne Robinson's much acclaimed 2004 novel. Jack needs the blessing. Having left home, followed a criminal path, abandoned one child and failed another, he is lost and loses much of what is dear to him. An outcast without direction, Jack moves (in) the story restlessly (*until he becomes its hero*). My analysis of *Gilead* begins with Jack, his head between Reverend John Ames' hands, being blessed. This is the culmination – but also the heart – of the story. "I'd have gone through seminary and ordination and all the years intervening *for that one moment*" (276, emphasis mine) says John Ames blessing his godson Jack. In a sense, this is true for us as readers as well. *Gilead* is John Ames' long letter to his own child;[4] a recounting of his life which we follow with each page for "that one moment," Jack's blessing. It is then that Ames forgives his godson's past – something the minister has grappled with throughout the entire novel – and hallows the possibility for Jack's reformed future. The blessing is, in fact, not the end but the beginning.

We know that *Gilead* will conclude with Ames dying. The letter is a parting gift to his son. "If you're a grown man when you read this," Ames tells him, "I'll have been gone a long time" (4). But the story does not end with John Ames. On the contrary, it anticipates the time after Ames' death, a future that will unfold in his absence (see Tanner). What follows is life, not death: Ames' son will grow to be a man; Ames' godson, Jack, has been blessed so, in a religious sense, will be living

---

4 In *Home*, we learn that Ames has named his son Robert Boughton Ames, i. e. named the child after his best friend Robert Boughton

as a "new" man, too. We do not know the details yet, but we can feel that this future story is there. We have been prepared for it early on. Ames' father has also bequeathed a letter and although Ames has burned it (7), generational lines are raised from the ashes as Ames recounts his grandfather's and father's stories and passes them forward to his son. The son, too, "might think of writing some sort of account [as an] old man" (239) tells us Ames, and anticipates his own re-emergence, "imperishable" and "more alive" than ever (60) in his son's text, after death. While *Gilead* recounts Ames' life, however, it is a story about the future that his past and present pull towards the characters. It is this future I want to consider.

My readers may wonder why, in talking about *Gilead*'s future, I do not engage Robinson's later novels, *Home* (2008) and *Lila* (2014). After all, these novels pick up *Gilead*'s characters and deepen its narrative world. As Rachel Sykes notes, however, *Home* and *Lila* do not extend *Gilead*'s chronology; they do not form a sequential trilogy but are "simultaneous fictions" (55). Indeed, Robinson's later works change narrators and help us see more of *Gilead*, but they remain part of the same picture. As Sykes notes, the image Robinson draws contains "an untold number of partner novels [with] parallel histories of the region ... extending the reader's knowledge of one narrative moment" (62). The "future" of *Gilead*'s characters then is not be found in later texts. In fact, their future is present all along in the novel.[5]

For Laura Tanner, *Gilead* anticipates the time after John Ames dies. The prospect of Ames looking back from the grave, Tanner explains, heightens his "appreciation of the present [and] the ability to move toward a reimagined future" (251). In this sense, the future shapes Ames' past and present and hints at its own emergence. Unlike Tanner, I suggest that the future is not only impending but buried in the text; it is already a given. The future, I argue, is more than a loss (Ames' absence) and more than enrichment to the present (the future's own absence), but a living temporality (or, temporality for the living) that already unravels in the text. Put otherwise, Robinson not only instils "the one narrative moment," the year 1956 in Sykes' analysis, with the time after it, but visualizes this time. In this sense, the future is not only approaching, but takes place; it awaits on the horizon but is also horizontal – a "point on the circumference of an enormous circle" (Smith 29).

This becomes clear if we examine how Robinson constructed *Gilead* as a story about one and the same man (or in Robinson's wording – a story about the "self-

---

5 Although his analysis focuses on the present as "experience of the divine in the immediate and the immanent" (349), Christopher Leise has also suggested that *Gilead* describes three temporalities – past, present, and future.

same" man [239]). In the following, I propose that the main hero in the story, "the beloved son and brother and husband and father" (276), is reflected in the image of all white male characters. I consider how these characters share appearance and experiences and repeat their hi/stories, and I investigate how such repetition marks them as reflections of each other, "blessed and broken" into different versions of themselves (79). I argue that walking in each other's footsteps (and into each other's physical form), the main characters not only continue lineage but demonstrate the imperishability of the future as 1) white fatherly property (Ames' heritable "begats" [10, 85]), and 2) as already begotten (Ames' future self(s) inhabiting the world).

Examining how the story portrays what turns out to be one ever-present and ever-progressing white Hero,[6] I underscore the relation between Robinson's representation of time (or, how the selfsame man experiences past, present, and future) and the way Robinson bars Black being and liveability (Pak 2015). My interest lies in Robinson's conceptualization of futurity and the ways this conceptualization intertwines with white ideology. For, the liveable/lived future in the story saves white male heroes and absolves their culpability (whatever their crime, it is "worn and stale" and already forgiven [139]). Contrary to this liveable/lived future in the text there surfaces an image of stifled and hopeless Blackness.

My first aim is to show how *Gilead* promises a brighter tomorrow to the Hero, and simultaneously poses this brighter tomorrow as an always already fulfilled destiny, a proof in and of itself. It is thus that the anti-Blackness underlaying *Gilead* (Pak 2015) becomes forgiven and white men are reconfigured as rightful and righteous harbingers of better times (*or, as dominant ideology would have it, they are reconciled with their destiny: again 'Redemption is the narrative inheritance of Humans'*). Blackness is used and jettisoned in the process, but this transgression appears pardoned in the larger context of the story. In other words, while Robinson exploits and injures Blackness by deeming it atemporal, she sets up exploitation and injury (and anything white author/ity does) as a minor transgression. My second objective is then to examine how the morality and hope read into *Gilead* relate to the ways Robinson attaches futurity to white bodies and dislocates it from Black bodies, and whether and how such white construction of futurity elicits a "white" reading, i. e. seeing exclusionary futures as something universally positive.

In the following, I discuss how each of the white men in *Gilead* resembles the others; how they appear as reflections in an infinity mirror: each one slightly different due to perspective but ultimately summable as one and the same, the "self-

---

6 I capitalize "Hero" to indicate a summation of all white male characters, i. e. I read their characterization together, as one image.

same" Hero. To avoid the vertigo of repetition, I pause on Jack and analyse how each of the other men compares to him. The collage of images creating the Hero enables the progress of (hi)story: we will see that the Hero is born and re-born continuously; he grows, evolves, and albeit mortal, he re-appears yet again. This is one of the ways futurity is invested in the Hero's body.

The other is Robinson's rendering of the future as imaginable: the future is visualized, named and known. As Yumi Pak argues, *Gilead*'s characters are divided by the way they experience time: white heroes participate in progressive history while Black characters do not ("Jack Boughton"). I follow Pak in relating white and Black time in the story,[7] and further Pak's arguments by exploring 1) how anti-Blackness is framed as forgivable (or, how the story motivates white hope/future time), and 2) how Robinson rewrites Toni Morrison's *Beloved* working against its critique of white hope and claim on being. I thus turn to the future promised by *Gilead* and consider how the novel sets up a context for celebrating white progressive hi/stories.

Jack the broken, Jack the blessed

Jack could easily be an Acker's character. Like Acker's rebels, he goes through and causes trouble but thus articulates social critique. As Jonathan Lear suggests, Jack's heroism lies in risking his life to expose a hypocrisy – Gilead's false morality (*Wisdom*). Clearly, Ames' blessing sounds disorientating in this context. While the blessing is given as forgiveness for Jack's transgressions, it also indicates the belated recognition of Jack's intrinsic goodness (*Jack's transgressions must have not been so terrible after all*). The fact that Jack is both a disappointment and a darling (82–83) baffles Ames all along but turns out to be the moral of the story (see Hinojosa, Weele). Comparing Jack to another famous misfit,[8] Suzan Petit notes that Robinson directs Jack's defiance and devilishness towards (his) greater good:

> Robinson ... wants her readers to see Jack's goodness, "the greatest goodness [being] perhaps the awareness of one's own failure to be good" ... a failure which is inevitable because of man's fallen condition ... Robinson clarifies how she sees Jack by saying ... "I'm fond of him. I didn't want to make Jack a good man in a conventional sense. I wanted to make him a person of value in terms of the whole complexity of his life: [I] wanted to see him redeemed." (302)

---

7 time experienced by white and Black characters, respectively
8 Petit compares Jack to the Misfit in Flannery O'Conner's "Good Man"

Like Acker's Kathy, Jack resists normative order and proves to be a controversial hero. Jonathan Lear goes as far as to read him as "a sinister, perhaps even diabolical figure" (269), but later retracts the dismissal:

> Jack … is doing God's work [he is] a Christian hero … This, of course, does not make Jack a virtuous man. He is not, and that is not the kind of heroism I am talking about. He is rather a challenge, an occasion for others either to display or fail to display their Christian charity. (280)

It is Jack, the misfit, who will make *Gilead* "a novelized treatise on the difficulty of lived virtue" (Painter 95). This interpretation in critical analyses follows directly from Ames' judgment of Jack. At first, Ames sees him as "the lost sheep, the lost coin. The prodigal son" (83). Albeit reluctantly, Ames admits that Jack disturbs the peace with his return, that Jack forebodes danger. While the disclosure of Jack's sin is postponed, however, readers are carefully prepared for it. On the one hand, Ames highlights Jack's childhood mischief. "I don't know how one boy could have caused so much disappointment without ever giving any grounds for hope" exclaims Ames (82), and goes on to list Jack's offences as a child: small thefts, setting things on fire, joyriding, and generally disgracing his family (206–209). On the other hand, Ames communicates his discomfort at Jack's presence in *Gilead:*

> That paragraph would itself amount to a warning. Perhaps I can say to your mother only that much. [Jack] is not a man of the highest character. Be wary of him. If he continues to come around, I believe I'll do that. (143)

"My evil old heart rose within me," admits Ames looking at his son and second wife, Lila, sat next to Jack, and pinpoints the nature of his fear: "leaving [his] wife and child unknowingly in the sway of a man of extremely questionable character" (160). Ames stalls, and neither his son, nor we, the readers, learn what is it that makes Jack dangerous; not until later. The context is set, however, and we, like Ames' son and wife, are warned: there is something to be distrusted in Jack. We await its articulation but know it to be there: Jack has sinned enough to shame and hurt his family, and to make his godfather uneasy. Deficiency is the first stroke in Jack's portrait: Jack has not lived up to his family name (Boughton is "a saint" [223], his sister rightfully named Glory) and to Christian morals.

The second stroke is Jack's blessedness. We learn that Jack is not as broken as Ames suspects him to be (*there is goodness in Jack which must have been there all along*), and that Jack moves away from sin and towards redemption (*goodness will be arrived at*). On the one hand, Ames admits his erroneous "habit of seeing meanness at the root of everything" Jack does (263) and stands corrected. Ames realizes Jack's inner beauty (265) and surprised remarks that it has only required "a little willingness to [be seen]" (280). At the end of the novel, Ames recognizes that Jack has returned to Gilead not to disturb the peace but to find it. Having fallen in love with Della, a Black woman, and fathered a son (247), Jack hopes that a "shining star of radicalism" like Gilead (250) might become their home:

> I came here, thinking I might find some way to live with my family here, I mean my wife and son. I have even thought it might be a pleasure to introduce Robert to my father. I would like him to know that I finally have something to be proud of. (261)

It is important to note that Jack is proud not only of his child but also of himself. As we have learnt a little earlier, the sin Ames defers telling is Jack's abandonment of his first born, a daughter (177–82). Jack does not marry the girl's mother – she is too poor and young and so Jack has "no business in the world involving himself with [her]" (178) – and the daughter dies soon after. Having failed this woman and child, Jack now appears as repentant and transformed. He asks Ames: "What about this town ... could we live here? Would people leave us alone?" (264). The new family turns out to be a two-fold blessing. On the one hand, this is Jack's chance to redeem himself and prove worthy as a father and husband. On the other, Jack's love for a Black woman might redeem his father's, and Gilead's, failure to uphold "Christian" values. Ames admits:

> I don't know how old Boughton would take all this [the news about Jack's Black wife and son]. It surprised me to realize that. I think it is an issue we never discussed in all our years of discussing everything. It just didn't come up. (251)

That Ames cannot guess how Boughton will take the news is, of course, not true. In 1956 America even a small town like Gilead sees its Black church burn and Black people disappear (264). Segregation laws make Jack's marriage with Della illegal, and he anticipates the obvious: even once an abolitionist hub, Gilead is not safe for Della and his son. It is also not true that Jack's father, Boughton, never considers the issue. In a conversation with Jack we read in *Home*, Boughton remarks that Black people only "bring [trouble] on themselves." Asked whether he has heard of Emmett Till, Boughton remembers Till as the "N-fellow [who] attacked the white woman" (163). Jack corrects him: Emmett Till is a Black child who was murdered; the attack was on **him**, not **his.** Disagreeing, Boughton concludes: "I think there must have been more to it, Jack," and swiftly shifts the blame from Till to Till's parents: "[t]hey bring children into a dangerous world, and they should do what they have to keep them safe" (163). Jack's desire to have a Black family seems to redress Boughton's racism.

Accepting Jack's new wife and child could be Boughton and Gilead's chance to atone, to be the good people in a good town they claim to be. After all, Gilead was once "a place [abolitionists] could fall back on when they needed to heal and rest" (Gilead 267), and Boughton an example of charity and goodness (193). As Lear points out, Gilead's people are blessed by Jack's return and are given opportunity to test and prove their charity. Recognizing the goodness in Jack and the grace he might bring with himself, Ames feels finally ready to love and understand him (276). He exclaims:

> I do wish Boughton could have seen how his boy received his benediction, how he bowed his head. If I told him, if he understood, he would have been jealous to have seen it, jealous to have been the one who bestowed the blessing. It is almost as if I felt his hand on my hand. Well, I can imagine him beyond the world, looking back at me with an amazement of realization – "This is why we have lived this life!" (277)

We do not know which epiphany Ames means here: that Jack is intrinsically good and capable of transformation, or that by accepting him and his family Gilead might restore its goodness. Bowing his head, Jack repents and improves. While the benediction marks Ames recognition of Jack's true character – not questionable but good – it also reveals Ames' own capacity to accept grace. Ames can finally see past the disappointment of both Jack and Gilead: the boy has proven to be "a person of virtue" and the town is no longer hypocritical. Ames concludes: "I love this town. I think sometimes of getting into the ground here as a last wild gesture of love" (282). As Vander Weele notes, Gilead's blessed are not only those who repent but those who bless them and recognize the existence of shared flaws ("The Difficult Gift"). The blessing thus entails hope that Jack might reunite with his family and the people of Gilead might return to their ancestors, to previous radicalism and readiness to help.

Being broken and being blessed are thus interwoven in Jack's character; being both at the same time, he personifies "the difficulty of lived virtue." Yet, Jack is also woven into Gilead's blessedness and brokenness. While he is lost and broken, Gilead seems like a blessed place; when Jack reveals his inner beauty, the town shows itself as "awkward and provincial and ridiculous" (267). At the end of the story, Jack and Gilead's images become interdependent: each is broken and blessed in relation to the other; each is in equal portion disappointment and grace. This interdependence is important to keep in mind because Jack's individualism combines all male characters in the story. A variation of Gilead's blessed and fallen men, every son and father in the story relives Jack's experiences. While their reactions and choices differ, however, the structure of existence and being remains the same: [9]

---

9 This coda does not hold true for the Black characters in *Gilead*

> There are two occasions when the sacred beauty of Creation becomes dazzlingly apparent, and they occur together. One is when we feel our mortal insufficiency to the world (*men are broken/fallen*), and the other is when we feel the world's mortal insufficiency to us (*men are good/blessed*). (280)

I have reviewed Jack's image in detail because his profile recurs in all white men in Robinson's novel, and it is necessary for considering my broader interest: how does the Hero inherit and beget time, what is the characters' relationship to the future, and in what ways does this relationship effect imagination and interpretation? As Lear suggests, Robinson puts the readers' own virtue and hopes on trial with a character like Jack (284–5). In my view, Robinson does something more – she visualizes and defines the extent to which virtue/hope can be lived; she directs us where to look for it and promises it as a future realization for white characters. This is an old Christian principle only as far as one forgets that Robinson's Christianity is conveniently censored and premised on anti-Blackness (see C. Douglas, Pak).

I follow Pak and Douglas here in suggesting that, rather than Christianity, the organizing principle of Jack's representation is dominant ideology. As we will see, in *Gilead* whiteness saves white heroes just as they save whiteness in return; the moral of the story is white heroism just as white heroism attests to white morality. Before I discuss this, however, let us see how *Gilead*'s white men are synchronized into one, the "selfsame" Hero, and how he will become an evidence for his own return (*return as a double entendre: recurrence/homecoming, and reward/income*)

The persistence of repetition

It is easy to imagine how the male characters in *Gilead* align with and overlay each other. One the one hand, they slip in and out of being the same man. For instance, male names and vocations are heritable: John Ames is the son of John Ames and the grandson of John Ames (10) and Boughton calls Jack after him as well (99). With the exception of Jack and Ames' brother Edward, several generations of fathers and sons are preachers (7) and pass down the same knowledge (141). Although it is a common tradition to bequeath one's name and trade, men in *Gilead* share something more than that. According to Jack, fathers also bestow identity (192) and do so not simply as a matter of lineage. Admitting his failure to accept Jack, Ames writes: "John Ames Boughton is my son" but quickly clarifies: "[b]y "my son" I mean another self, a more cherished self" (215). Here, Jack stands for both Ames and his son. Similarly, Jack calls his godfather "Papa" and his godfather's son "brother" (104–5) and introduces himself with Ames' name and takes his place (104, 160).

The spitting image of his own father (106), Jack is both Boughton's son and his incarnation. "[Taking] after his grandfather and father" (161), Jack carries the features of several generations. Yet, Ames also dreams of being Boughton (162) and imagines himself as his own grandfather (100). Ames' grandfather is in turn mistaken for Jack's grandfather (259). On the other hand, male characters experience rather than simply exhibit similarity. For the sake of brevity and because I am not concerned with plot twists, I have divided parallel experiences according to *Gilead*'s motifs and themes. The following ten sections illustrate dominant repetitions which structure white male heroes' experience in *Gilead* and thus enable the vision of white men as one and the same.

1 Disappointment and offence

Gilead's men fail and insult their fathers. As mentioned earlier, Jack is a boy who continuously disappoints Boughton and Ames (82). Like Jack, Ames and his brother Edward disappoint their father (8, 30) who has disappointed their grandfather (11, 39, 95–6). Furthermore, Ames' father is offended when Ames' grandfather preaches (200) as is Jack when Ames does (148, 193). In turn, Jack offends Ames (197); Boughton's father is also offended (134).

2 Family

Like Jack, Ames has a previous child, a daughter. Just like Jack's first child, Ames' daughter dies (20). Although Jack abandons his first family and Ames loses his unwillingly, both occasions are paralleled. Thus, Ames feels the shame that Jack should feel thinking about his dead daughter (23). Ames and Jack's second families are also analogous: they are both an "object of scandal," "unconventional" (262); neither Jack nor Ames see their sons grow old (62). Moreover, Jack takes Ames' place in Ames' family (149, 160), Ames wishes to give his wife and child to Jack compensating for the loss of his wife and child (266), just like Boughton "gives" Jack to Ames compensating for the fact that Ames does not yet have a family (99). Like Jack's first wife (178), Ames' second wife is too young (262); love leaves both Ames and Jack sick (235, 274).

3 Words and sermons

Some words burn. Glory warns Jack that words might kill Boughton (249), just like Ames' mother warns Ames that they might kill his father (31). Sermons are repeatedly written and burnt (7, 280), and are heritable. Ames learns his sermons from his father (21) who has written a letter to him (7) like the one Ames writes to his son. The son might, in turn, write a letter (239).

4 Crime and transgression

Like Jack's stealing as a child (206 – 9), Ames and his father's thefts only faintly resemble crime (18); Ames' grandfather is also compared to a thief but is not really one (35).

Just like Ames senses that Jack has done something wrong, he feels that his grandfather has, and remarks that all men are responsible for the crimes whether they know about them or not (93, 215). While somebody shoots at Ames and his father (17), Ames' grandfather shoots somebody else (123).

5 Catch

The players alternate. Ames plays catch with his father (20,72), Jack plays with Ames' son (115, 135), Boughton plays with Ames (116), and Ames plays with Edward (131).

6 Miracle/darkness

Like for Jack, the metaphorical period of darkness in Ames' life leads to the miracle of having a family (63, 271). Having a wife and child is equally surprising to Jack and Ames (59, 247); they both return to darkness when they lose metaphorically sight (of their families) (64).

7 Ambiguous character

Men's character is exaggerated and misunderstood. Jack is not as good as his family thinks (82) and not as fallen either (264); Ames is not as good as Gilead's people believe (74), neither are Ames" grandfather (93), brother (29), son (60), and Boughton (251). Several generations of men are angry (7, 11, 39), saint-like (45, 36), and afflicted (56, 50, 274).

8 Death

When Jack returns to Gilead he finds his father on his death bed. Jack then goes away, leaving his father to die without him (274) just like Ames and his father have done (11, 19, 268). All men in the story die old (5, 134, 279).

9 Church

As a child, Ames helps his father to pull down a burnt church (107), Ames knows that his own church will be pulled down (80); Ames' grandfather leaves Gilead soon after the Black church is burnt (41–2).

10 Journey

Jack's journey turns out to be a blessing to him and Gilead (276), Ames' journey with his father turns out to be a blessing as well (20). Like Jack, Ames' grandfather, father, and brother leave Gilead (41, 268).

On the one hand, these parallels confirm a familiar dictum: history repeats itself, there is "nothing new under the sun" (Ecclesiastes 1:9), all experience is "worn and stale" (*Gilead* 139). What happens to Jack and Ames might happen to Ames' son, just as what has happened to Ames' grandfather has already transpired in someone else's life. On the other hand, congruent occurrences frame each man as a version of the other. We see Ames as a young boy and an old man resembling his young son as an old man (239). Braced by repetition, the representation works towards synthesis: men appear as each other, and as one and the same. Observing thousands of fireflies, all of them identical and all joining into one fire, Ames remarks: "I believe the same metaphor may describe the human individual (...) Perhaps Gilead. Perhaps civilization" (82). Later, Ames tells us that the light of fire is "constant [and people just] turn over in it" (239). Turning, they sometimes flash as blessed, sometimes as broken, but are always both – even when the duality remains unseen. Thus, Ames' letter, *Gilead*, is not simply his way of "considering things" but an attempt to capture the duality. Jack's blessing is worth Ames' long letter because it illuminates true essence. Ames tells his son: Jack is "a man about whom you may never hear one good word, and I just don't know another way to let you see the beauty there is in him" (265).

The effect of conflating men into one and the same Hero is a reassuring con-statation for white readers. Even if Ames dies, the story will go on. Or, as Arthur Rimbaud upliftingly asserts, other men "will come [and] will begin at the horizons where the first one has fallen" (qtd. in Nielsen 35). Yet, future is not only promised but seen in Gilead. The future self of each man appears in the text because 1) the Hero turns from a boy to a dying man to a boy again; and 2) Ames visualizes his future. He writes to his son:

> Why do I love the thought of you old? That first twinge of arthritis in your knee is a thing I imagine with all the tenderness I felt when you showed me your loose tooth. Be diligent in your prayers, old man. (239)

Moreover, the text reveals what happens after the last page. For instance, Ames knows that he will vote for Eisenhower (107), and that when he dies his church will be pulled down (80), his family will have a hard life (5), his son will leave Gi-lead (281), and new wars will come (50). To a similar effect works the moment when Ames writes his funeral sermon and imagines Boughton reading it (140). Fu-ture time is thus not only anticipated but visualized. It is both approaching and already before us. Reading Ames' letter – in the future – his son emerges simulta-neously as a boy and an old man. Ames tells him: "you have grown up (...) you poor child, lying on your belly now" (118).

Arguably, Robinson derives the ever-present and ever-progressing condition of being from Christian doctrine.[10] As a religious writer, she borrows extensively from Scripture where we read that a) "we all ... are being transformed into the same image" (2 Corinthians 3:18) and b) this image lives on as each "generation goes, and a [new] generation comes" (Ecclesiastes 1:4). Within Christianity, genealogy and time are both linear/vertical (fathers beget sons who beget sons (b)) and, as Smith suggests ("The Nature"), horizontal (fathers and sons are one; they are self-same/simultaneous) (a)). Whatever *Gilead*'s logic and logistics of time, however, they do not hold for every character in the story. As Yumi Pak argues, time in *Gi-lead* is split: white characters progress while Black characters do not, and the fact of the latter enables the fact of the former:

---

10 See for instance Ellwood Johnson's *The Goodly Word: The Puritan Influence in American Liter-ature*

the physical violence of eradicating Blackness, coupled with the psychic violence of expropriating Blackness as a ghostly presence to be used at will, is what [allows] for the continuation of white genealogy. (214)

There are two conclusions we may draw here: 1) it is Christian time that excludes Black characters; or, differently racialized people inhabit existence/being differently, and 2) time in the story is organized through another principle altogether. In all my atheism, I concede that to writers like Robinson, the first inference would sound self-cancelling.[11] To consider the second, I now turn to Robinson's construction of Black-~~time~~.

Rewording

While Robinson constructs time for the Hero via progressive repetition (fathers beget sons who beget fathers who beget sons), she uses what I call "rewording" in excluding Blackness from this time. On the one hand, "rewording" points to the textual instrumentalization of Blackness: the ways Blackness is purposefully verbalized and unspoken (put into words and deleted). Following Toni Morrison, Yumi Pak argues that Black characters in *Gilead* are continuously brought in and out of utterance and are thus used to sustain narrative coherence ("Jack Boughton"). Put otherwise, *Gilead*'s author/ity renders Blackness "raw material" (to go back to Hannah Black) which is then conjugated for the purpose and production of meaning. As Black critique shows, however, Blackness is always already coerced to mean *insert dominant desire here* (see Fanon, Spillers, Wilderson). Meaning is then simultaneously thrust into and subtracted from Blackness. I take "rewording" to demarcate both directions of this process. On the other hand, "rewording" refers to Robinson's intertextual gestures. In *Gilead*, she revises a famous story and history as experienced by Black, rather than white, characters: Toni Morrison's *Beloved*. My analysis will show that the changes Robinson makes to *Beloved*'s plot 1) influence directly how we perceive white and Black time in *Gilead*, and 2) enable the opposition of *Gilead*'s conceptualization of time against that in *Beloved*. I will first discuss how Morrison's text appears in *Gilead*, and then consider how Robinson redeploys and adjusts it.

---

11 there is a heated debate over the characterization of Black bodies in the Bible (McKissic, Felder, Mbiti, Kealy, Shenk) and over what Frederick Douglass exposed as Christian racism. However, I have not found any evidence that Robinson considers Christianity as exclusive or racist. On the contrary, Robinson underscores that there is is a universalism underlaying all human experience, some "ultimate truth" about this experience as non-exclusionary ("The Faith Behind the Fiction")

Beloved tells the story of an enslaved woman, Sethe, who flees the Sweet Home plantation with her children. They settle at 124 Bluestone Road – a house in Ohio where Sethe's mother-in- law, Baby Suggs, lives after her son has bought her out of slavery. Schoolteacher, who has taken control of Sweet Home after the death of the previous legal owner, traces Sethe and her children and plans to bring them back to the plantation. Realizing that her children will be forced back into slavery, Sethe tries to kill them and herself; death being the better alternative – "death [as] a synonym for sanctuary" – to Sweet Home's horrors (Wilderson, *Afropessimism* 243). Sethe manages to cut the throat only of her two-year-old daughter. The child dies in Sethe's hands. Seeing this and deeming Sethe too crazy to be useful, Schoolteacher changes his mind and leaves without her and the children. Sethe is then sent to jail together with her surviving daughter, Denver. They return to 124 to find the house haunted by the ghost of Sethe's murdered child. We never learn her name but read her as Beloved – a word engraved on her tombstone, the only one Sethe could afford. Sethe's other children – boys named Howard and Buglar – run away and Baby Suggs dies, leaving Sethe and Denver alone in the haunted house. Paul D, a slave from *Sweet Home*, arrives and chases the ghost away. For a while, it seems that Sethe, Paul D, and Denver can live like a family. Crawling out of a river, Beloved rises from the dead as a young woman and moves in with them. Beloved seduces Paul D, Sethe tells him the details of Beloved's death, and he leaves. Sethe and Beloved become obsessed with each other, Sethe quits her job to stay at home with her girls. She grows weaker while Beloved turns more and more insatiable, big-bellied and expecting (care or a child or both). Scared of losing her mother, Denver finally gathers courage to leave the house and looks for help. She finds work as a maid for a white family. When the white employer approaches 124 to take Denver to work, Sethe mistakes him for Schoolteacher and tries to stab him. Present at the scene are local Black women who, no longer resentful of Sethe, have gathered to drive the ghost out. Beloved disappears. Paul D returns promising to take care of Sethe; Denver is no longer held and hidden in 124, and following the promise of a more inclusive future, steps out into the world.

Many of *Beloved*'s themes and expressions re-appear in the description of Jack's first child, his daughter. We learn what happened to her in four brief pages. Twenty years earlier, Jack seduces and abandons her mother – a young, poor girl living near his university. Jack does not care for either the girl or the daughter. He confesses what has happened to his family and the Boughtons start visiting the girl's home, wanting to care for the child and even take her with them. The mother refuses to let her go. When the child is three years old, she cuts her foot and dies of the infection. He mother runs off to Chicago and the Boughtons never hear of her again. Below, I list the major similarities between the daughter's story and *Beloved.*

Daughter and language

Like Beloved, Jack's daughter is a **"sullen pride** to her mother" (*G* 181). In *Beloved*, both mothers and daughters are **"sullen"** (*B* 47, 121) and their love **"prideful"** (*B* 137, 256, 272).

Like Beloved, Jack's daughter **dies from a cut** (and because her mother cannot protect her) and **before she is given a name**; Like Beloved, she is about to be **stolen** (*G* 181); **her mother also runs away** (*G* 182).

Like Beloved, **the dead girl** (not Jack's daughter but her parallel – Ames' daughter) **comes back** and is **full of reproach** (*G* 23).

Like Beloved, Jack's daughter is the reason a woman is (potentially) **imprisoned** (*G* 181). Glory, Jack's sister, wants to **steal his child** but knows that if she does, she will go to **jail**, and the daughter will be returned to her mother. In *Beloved*, Sethe's escape amounts to **theft** (in slavery, running away means stealing the legal owner's property; Beloved returns to her mother, who has also been to **jail.**

Like Beloved, Jack's daughter is seen in the **river.** Sethe plays with Beloved on a **frozen lake** (B 174), Beloved comes out of **the river** (B 50) and is seen to stand in it (B 105). In Gilead, the mother and daughter are seen standing and playing in **the river** (*G* 185).

Like *Beloved*, *Gilead* connects family love to **greed and guilt:** while Beloved is **insatiable** and uses Sethe's **guilt** to demand more and **more food and attention** (*B* 250), in *Gilead* the family blackmails Jack's father and wants **more money** taking advantage of the **guilt** he feels (*G* 181).

Morrison's famous lines **"[i]t was not a story to pass on"** and **"[t]his is not a story to pass on"** (*B* 275) appear in **Gilead** as **"[i]t is a bitter story"** (181).

As Orlando Patterson points out, the system of slavery functioned through severing kinship among the enslaved (*Slavery*). Dehumanized and natally alienated, children of enslaved parents did not belong to them, neither did these parents belong to each other and to their own families. In *Beloved*, Morrison writes:

> In all of Baby's life, as well as Sethe's own, men and women were moved around like checkers. Anybody Baby Suggs knew, let alone loved, who hadn't run off or been hanged, got rented out, loaned out, bought up, brought back, stored up, mortgaged, won, stolen or seized. So Baby's eight children had six fathers. What she called the nastiness of life was the shock she received upon learning that nobody stopped playing checkers just because the pieces included her children. (23)

Note: emphasis mine throughout; *G* stands for *Gilead*, *B* for *Beloved*

Enslaved, Sethe could never truly mother her daughter; she had merely produced property (which Schoolteacher claims as such). Robinson picks up this vocabulary: Jack's **daughter is produced**, rather than simply born: Jack's "involvement [with the girl] **produced a child**" (177); Jack also does not acknowledge his child – like slave owners who often fathered children with enslaved women but never considered them theirs.

The house, the tombstone, and the kidnapping

Like 124, the house where Jack's daughter and her mother live is "**isolated**," "**sad**," and "**hostile**" (*G* 178 – 9); Jack's daughter is kept "**a hostage**" in it – a fact reminiscent of **Beloved**'s famous lines "**she is mine**" – "**you are mine**" (*B* 210 – 217), spoken in turns by Beloved, Denver, and Sethe (**mother and daughters keep each other hostage and are also hostages in 124**).

Like Beloved's **tombstone**, that of Jack's daughter is not engraved with her name; instead the tombstone reads "**Baby**" (*G* 181).

Planning to **kidnap** Jack's daughter, Glory models the plan on how **abolitionists** smuggled slaves: Glory knows "about the old days when they used to **smuggle fugitives** up from Missouri, and she [thinks] one small infant would be a much easier thing to conceal" (*G* 180). In *Beloved*, a white woman helps Sethe and her daughter (who are **fugitives**) to reach Ohio. Later, women gather to **exorcise the ghost**; in *Gilead*, Glory also calls a doctor to the rescue (*G* 181).

Taken on its own, each of the above parallels amounts to nothing further than a chance resemblance. After all, vocabularies are limited and highlighting reoccurring words in any two texts as I do above would easily trigger Déjà-vus: a feeling that *Beloved* emerges in *Gilead* whether it does or not. The choice of symbols in both stories is also logical: death is traditionally linked to tombstones; afterlife reunion between loved ones is a common consideration, and so on. However, all resemblances to Beloved spread on the tightest of spaces – 4 pages in Gilead, the brief recounting of up-to-this-point carefully deferred information. All resemblances appear together – and stronger for it, rammed and fixed as they are – the nameless dead girl, the engraved tombstone, the child kept captive in an isolated and hostile house, the deadly cut, etc. Moreover, the concurrences are paralleled to slavery and enslaved people escaping the plantation, and thus brought closer to *Beloved*'s setting.

Can we speculate and say that four pages in *Gilead* are haunted by Morrison's text? Literary critics have already suggested that *Gilead* is haunted by Black presence (Bailey, C. Douglas, Pak) – a circumstance that would certainly support the interpretation. Furthermore, intertextual plays are not foreign to Robinson's style.[12] For me, the commonalities between both stories are too many and too close to disallow voluntary **[jumpcuts]:** Beloved pushing under the ribcage of a carefully kept pronunciation, Jack's past sin, four pages of (hi)story. Yet, the stories are fundamentally different. Whereas *Beloved* thematizes the "always now" of Black suffering (see Childs, Sharpe, Wilderson, Murillo) and returns life to bodies treated and imagined as unliving, its reworded version skips past it. In Robinson's story, Black life is not portrayed as in danger but deadly, it is considered a danger in and of itself.

Before I address this issue, I need to make one important clarification – Jack's daughter is not explicitly racialized as Black, neither is her family on her mother's side. Yet, the family is othered, and approximated to Black Positionality. This is neither a unique, nor unlikely manoeuvre. The history of conflating poverty and Blackness, the so-called *race-coding of poverty*, is long and entrenched in US imaginary (see Clawson), as is the racist idea that Blackness equates with some and all kinds of deficiency (see Fanon). In the discussed story, poverty is used to mark the antagonistic relationship between J ack (and his family) and the girl (and her family). The latter live not only in "desolate, even squalid" conditions but outside the (normative) world. Ames remarks,

> How Jack Boughton even found [the girl] has never been clear to me ... It was a sad place and she was a sad child. And there he was with his college airs and his letter sweater and that Plymouth convertible ... Jack Boughton had no business in the world involving himself with that girl. It was something no honorable man would have done. [Jack] did tell his father about it. As if confessing a transgression. (178)

At first glance, it appears that it is poverty what distances Jack and the girl and makes their relationship dishonourable. This framing of poverty, however, is a contradiction in Ames' otherwise reliable narration. Throughout the novel, poverty is romanticized: being poor is "a matter of pride" and generosity, and poverty is "holy" (19, 35). The heroes in the story are also poor (5, 35, 226) but never in degrading, alienating ways. In contrast, the poverty of the girl's family means something else, and marks an irreconcilable distance between Jack's position and that of the girl. In fact, no amount of money can bridge the disparity between them. Thus, Jack's **daughter dies from a cut because she is barefoot** – the lack of shoes signifies the family's poverty – but **the child is barefoot not because she does not have shoes and her family money.** The Boughtons have already given them these: "We bought her shoes," Mrs Boughton said. "Why was she barefoot?" The girl said,

---

12 See for instance Alison Jack's "Barth's Reading of the Parable of the Prodigal Son in Marilynne Robinson's *Gilead*"

"Saving' 'em'"" (182). **The girl is poor (and dangerous) whether she is given money or not**; there is something chronic to her situation and fundamentally different to the ways other characters are poor in the story. With chronic poverty, Robinson does not denote the absence of financial means but a **condition and a lifestyle** incompatible with that of Jack and his family. The girl's poor judgement and behaviour cause actual death, and seem inevitable; although the Boughtons want to help, she is stuck to her ways and her kind.

The difference in positionality is further implied by framing Jack's relationship with the girl as "a transgression." We already know that being poor could not have accounted for this: the characters in the story "know how to be poor" and poverty is insignificant to their love and family life (35). Unlawful during *Gilead*'s narrative moment is not liaising outside one's class but outside one's race. In fact, Robinson does not obscure this historical detail but reminds us of it: Jack's relationship with Della occurs in the context of anti-miscegenation laws (261). As transgressive, Jack's affair with the girl is closer to relationships between white and Black people in segregationist America, and parallels Jack's later involvement with Della. It is, of course, true that Robinson never explicitly calls the girl and her family Black. Yet, she makes them inhabit a positionality contrasted to that of normative society.[13] This demarcation adds yet another referential gesture to *Beloved*. Based on my previous arguments, I make two important **[jump-cuts]** here: 1) I will analyse the story reading the girl and her daughter as Black*-end, and 2) I will discuss what happens to them in relation to *Beloved*.

From the perspective of Afropessimism, Sethe's murder is a salvage from the world, from anti-Blackness (Wilderosn, *Afropessimism* 243, 303). With the cut, Sethe reclaims herself, motherhood, and her daughter; Sethe does the unthinkable to protect her child, to demand another world and other ways of Being. The escape is needed because white people are coming for them, white people are everywhere, and they simply "don't know when to stop" (*Beloved* 104). Unlike Beloved, Jack's daughter dies for no reason: "she cut her foot somehow and died of the infection" (*Gilead* 181). The only danger for Jack's child is living with her mother, away from the better life the Boughtons promise.[14] The theft of the child, something Glory plans recalling abolitionists and fugitives, is rationalized not against but through

---

13 as we know from Afropessimism, Blackness is positioned in an antagonistic relationship to dominant discourse, i.e. normative, white society (see Fanon, Wilderson, Sexton)

14 as other anti-Black practices, the theft of Black children for the benefit of white families continues to this day. This theft is widely justified with the "better life" white families offer. Black and Indigenous activism challenges the theft and exposes its long-term consequences – see for instance the work by groups like *Grandmothers Against Removal*

white ideology. According to the latter, Blackness is the ultimate danger.[15] Within white ideology, Fanon explains, Black bodies are declared "impervious to ethics, representing not only the absence of values but the negation of values … the enemy of values … absolute evil. A corrosive element, destroying everything within its reach" (6). Fanon continues:

> colonialism was not seeking to be perceived as a sweet, kind-hearted mother who protects her child [the colonized] from hostile environment, but rather a mother who constantly prevents her basically perverse child from committing suicide or giving free rein to its malevolent instincts. The colonial mother is protecting the child from itself, from its ego, its psychology, its biology, and its ontological misfortune. (149)

Following colonialism's logic, Glory desires to protect the child from her own mother and lineage "whose ego, psychology, biology and ontological misfortune" will inevitably result in self-destruction (this is the same logic defended by Boughton when he blames Till and his parents for Till's murder). In agreement, Ames exclaims: "that was no place for an infant" and verbalizes the racism underlaying longstanding practices of stealing Indigenous and Black children from their families. Seeing time in poverty as dangerous, Glory dreams of having the child for even *"just one week!"* Here, "one week" is the temporal signifier for a time full of life (time with the Boughtons):

> [Glory] had washed the baby and dressed her up and sent [Jack] smiling photographs. She had photographed the baby in his father's arms … I suppose we really should have stolen her. (180–181)

Opposed to "one week" is the time with the mother – dangerous time; time that brings about death. This is not the same temporality to which Sethe sends Beloved. In Morrison's version, Black insistence on Being – Beloved – pushes from under the weight of the world, comes back, raging, and loves: "in this here place, we flesh; flesh that weeps, laughs; flesh that dances on bare feet in grass. Love it. Love it hard" (*Beloved* 88). Motherlove is strong enough to cut a throat but is motherlove nonetheless (*Beloved* 132). The bare feet in Robinson's version get cut, infected, and cause death: there is no reason given that the child should die but she does. "In those days you could die of almost anything, almost nothing," says Ames (*Gilead* 182). Just like Boughton has done thinking of Emmett Till, Ames immediately allocates criminality to the child's mother: she poses rather than prevents danger; she has not loved the feet of her own daughter enough to keep her safe (*Gilead* 182).

---

15 See also Christina Sharpe's discussion on the campaign "the most dangerous place for the African American Child is the Womb" in "Blackness, Sexuality, and Entertainment"

Two images emerge in such framing of the story: 1) time with the mother (Black time) is equated to unliveable life, or death, and 2) time with the Boughtons is equated to liveable life and future.

At the centre of the story rests Jack's sin: he has involved himself with a girl he should not have even met, and abandoned a child (178). The sin is crucial to Gilead because it contextualizes Jack's devilishness and later transformation *Again, "Re- demption is the narrative inheritance of Humans"* More importantly, the sin serves as a divide between two temporalities. It allows us to see what is good/fulfilled time, and what is dangerous/empty time. While white characters can go, see, and momentarily experience the latter (they can fall from grace, get off track), Black*end characters are locked and stuck. In the story, white men lose or waste their time but always return to their "senses" (and sensible/full time). Thus, Ames' brother Edward wastes time by studying to become a preacher but not becoming one (29). Jack wastes people's time by displacing their possessions and forcing the owners to search them for days (208–209); Ames wastes his time thinking/talking dishonestly (183, 192) and Jack loses his time in a protracted quest for the right path in life (276). Because time can be used (properly or not), it appears as full, as tangible enough to be expended or earned back. Although, the Hero often loses track of things – something evident in Ames' chaotic narration – he always comes back to (hi)story's due course. Edward reunites with his father despite his father's disappointment at Edward's atheism (267); Jack returns the sto- len objects (209), Ames recognizes the worth of talking and the true meaning of meeting Jack (276); and Jack finally repents and finds grace (275).

Contrary to this representation, operates empty/useless time: the time empty of life and prospects for the future. As mentioned earlier, Glory opposes the "one week" of good life to the time of poverty (time without life, Black time). Sim- ilarly, Jack's son and Della are suspended in repetition without hope; they move "back and forth" (260) without any prospect for resolution. Similar to the girl and her daughter, Della and her son are left stranded and lost. The narrative moves past them, as do other characters who find a way back to fulfilled time. The contrast between Black and white temporality thus reinforces each side of the antagonism (Pak). The more the first is adjourned, the faster the second re/ builds. The future plays a major role in this process – its negation establishes stasis, and its promise continuance.

In Robinson's rewording of *Beloved*, Black*end and lost characters are deemed too minor to return or haunt the story. In *Gilead*, the girl and her daughter do not appear past the four pages. They vanish giving space to Jack's protracted journey to redemption. Like the other Black characters, the girl and daughter are "narrative and symbolic material" (Pak 214). While they bring coherence to the novel – their brief story substantiates Jack's need and capacity to redeem himself – the girl and the daughter remain marginal characters. Such marginality further acquits Jack's sin. Brushing past the girl and daughter, the story goes on, and replaces them with another Black family. Even Ames, to whom Jack's sin had seemed unforgivable, changes his mind: "God bless the poor devil … I felt as if I'd have bequeathed him wife and child if I could to supply the loss of his own" (266). Here, by "lost family" Ames does not mean Jack's daughter and her mother but refers to his second wife and child, Della and her son. Although the girl and her daughter have disappeared, life and story continue; what Jack should have felt for them, he now feels for Della and her son. Mothers and children are substituted – if something happens and they vanish, others arrive in their place.[16] Thus, Ames and Jack each lose a wife and a daughter, but soon are blessed with another. What is lost appears as minor and finite, and allows the story to right "itself upward, in the direction of beauty and grace" (Row 114).

For white characters, the story ends well. Even Ames' expected death does not adjourn white time – he is imperishable in the image and imagination of his son (60). In contrast, Black characters do not extend to future generations: some of them disappear from *Gilead* (195, 264), others from the story (182), and those who survive cannot enter lineage and (hi)story. As Pak argues, Black characters secure "kinship between white fathers and sons" but remain irredeemably barred from it (224). Thus, Jack's son does not carry the family name and identity, nor can he cross the threshold of genealogy – the shared space available to white heroes in the story (Pak 233). As if to meet Blackness halfway, Jack leaves home and voluntarily strays away from his own (hi)story. Yet, this gesture does not unite Jack and his son. On the contrary, it marks a deep and incurable divide between them (Pak 221–223). Jack and his son are lost to each other; Jack's character moves too fast, answering to a different world order, and experiencing the world differently.

Rather than critique the divide – something, Afropessimistically,[17] Morrison does with *Beloved* – *Gilead* frames Jack's survival and return to White time as positive. For, while Jack (and his family as we see in *Home*) fail to reunite with Della and her son, he is framed as a character of "lived virtue" and as deserving of

---

16 I made a similar argument about the men in the story when I discussed the "selfsame" Hero. Yet, there is a fundamental difference between men and women taking each other's places. Unlike women and children in the story, men swap places and identities but always return – even when they are dead – to *fully* embody their own images. The substitute between Jack's first and second wife/child only erases their individuality; the replacement is complete: the first wife/child disappear as does the difference between them and the second wife/child. Later, the second wife/child disappear as well.

17 to borrow Paula von Gleich's use of the word here (*The Black Border* 11)

(Ames') blessing. As Robinson admits, she "wanted to see him redeemed" (Petit 302). By rendering *Gilead* a universal tale about hope and the beauty of human survival, literary criticism redoubles Robinson's gesture. Thus, *Gilead* is praised for inviting "the reader to participate in the act of transformation" and Robinson's oeuvre as redemptive and virtuous (Muhlestein, "Vision as Creation"). In this light, not only the story "rights itself up" but so does the morale of things getting better by simply staying the same. After *Gilead*, no transgression seems terrible enough to taint white heroes and white heroism. Lost in romanticization of the human experience, Black characters are used as reminders of what would happen if the Hero were to lose sight of its value. For, the danger Black characters are made to signify – "the negation of values" as Fanon reminds us – threatens not only the Hero's fleeting presence(s) but historical progress altogether. Or as Wilderson writes, "[h]istory and redemption are the weave of narrative... history and redemption (and therefore narrative itself) are inherently anti-Black" (*Afropessimism* 226).

5 Another town, another story

If you happened to walk in Pittsburgh, at Baum and Highland Ave in March 2018 you would have seen the letters "THERE ARE BLACK PEOPLE IN THE FUTURE" flashing. Posted on a 36-foot-long rooftop, the installation was designed by artist Alisha B. Wormsley. The sign derives from Wormsley's artwork[1] of the same name and was extended to the art initiative *The Last Billboard* – a project created by Jon Rubin and responsible for the tags on the East Liberty building at Baum and Highland since 2013. Wormsley's words were the project's last operation after "THERE ARE BLACK PEOPLE IN THE FUTURE" was taken down over objections to its content.

The company managing the property, represented by Eve Picker, demanded that the billboard be dismantled. According to Picker, people complained about the sign being "offensive and divisive" and as property manager she had to do right by the community, to take it down (O'Driscoll, "There Are"). Although previous billboards had also contradicted the lease – messages must be approved by the manager beforehand which they were not – Picker never challenged them. For, as she explains, "there has never been a community response about them being distasteful, offensive or political"; until then (Sostek, "Landlord"). Wormsley's sign was the first to receive negative "community response" and the subsequent ban. Condemning the removal of Wormsley's work, Rubin notes

> I find it tragically ironic, given East Liberty's history and recent gentrification, that a text by an African American artist affirming a place in the future for black people is seen as unacceptable in the present. (gtd. in Gordon, "Updated")

In the following, I will discuss whether and how Wormsley's message was reworded, more specifically whether 1) its meaning was changed to positive as suggested by Rubin's line of defence above, and 2) positivism hurts the critique inherent to a more "negative" aspect of Wormsley's work. Seen from the perspective of Afropessimism, rewording reveals how, after being subjected to a public trial, Wormsley is bound to lose whether she wins it or not. Or, how Wormsley's "winning," i.e. having the right to voice opinion, create art, and be present in public discourse, depends on framing her artwork as ineffective. As we will see, Picker concedes to reinstate the sign after its removal. This is the alleged win for Wormsley. However, Picker acts on the grounds of positivism re/turned to the message of Wormsley's billboard. Something gets lost in this process of rewording. In fact, Wormsley's cri-

---

1 see https://alishabwormsley.com/there-are-black-people-in-the-future-2

https://doi.org/10.1515/9783110799996-006

tique and claim on the future are defused. To understand why, it is crucial to keep in mind that Wormsley's message is not as positive as white audiences need it to be (to live with it). On the contrary, the premise of Wormsley's project is a critique of whiteness and of the violent removal of Black people from the future's landscape. Before we zoom in there, let us consider the reactions around the billboard's removal.

The racism buttressing the complaints against Wormsley's billboard has been widely discussed. "[M]y Pittsburgh has stood up!" writes Wormsley on her website as supporters protested the removal in social media. The Pittsburgh Office of Public Art also issued a statement in support of Wormsley, and a community forum was summoned to discuss the issue. Subsequently, Wormsley launched a new artwork-in-residency program giving chance to East Liberty's residents to work through and explore further "THERE ARE BLACK PEOPLE IN THE FUTURE."

Central to the narrative defending Wormsley's work was framing the billboard as positive, indeed antonymous to how it was first categorized. For Rubin, the billboard "[affirms] a place in the future for black people" (gtd. in Gordon, "Updated"). The Pittsburgh Office of Public Art echoes the same sentiment when they call the artwork "a positive affirmation" (Rose Sharpe, "Artist's Billboard"). This line of defence contradicts directly the racist assumption that Wormsley's message has somehow attacked and thus disconcerted certain residents of East Liberty. Indeed, the defence follows defence protocol: by disproving the first claim made about the billboard, namely that it is "offensive and divisive," the defence confirms Wormsley's innocence. Framing the billboard as positive was quickly picked up by Picker herself. Backtracking from her initial decision, she writes:

> Over the last 24 hours, we've received a number of emails from people who said they are not offended by the sign and are saddened by its removal. They far outnumber the people who originally approached us about being offended. We truly appreciate the comments from people who reached out to us in a respectful, thoughtful manner and believe the public has spoken. We are giving the tenant [Rubin] full approval to reinstate the original sign. (gtd. in Gordon, "Updated")

The (public) message Picker received is that Wormsley's billboard is not offensive; that in fact, people liked seeing it and would like to see again. Although well-intended, the defence re- framing the billboard as positive functions in several ways, and not all of them are positive for Wormsley. On the one hand, the defence includes an explanation (there are Black people in the future but do not worry, this is a positive message). On the other, it articulates Wormsley's message through a different authority (if Rubin, art institutions and the public say it, then it is O.K.) One could question why a simple sentence containing no obscene language required interpretation and power-ful echo in the first place. From the perspective of Afropessimism, this is a rhetorical question. Labelling the words of a Black woman offensive is consistent with the logic of the present moment because the world considers Black women's very Being an offence (Wilderson and P. Douglas, "The Violence"). Wormsley can say anything or nothing at all, white community is always already authorized to police her. "Black speech is always coerced speech," Wilderson reminds us ("We're Trying").

Yet, policing refers to something more than the removal of the billboard (i.e. the erasure of Wormsley's voice). While endorsing Wormsley's words as positive is surely a sign of support, at the heart of such endorsement lay the disarming Wormsley's body (of work). Claiming that the "f" word **Future** did not mean offence, the defence narrative insists that the message sounds positive, harmless. On the one hand, any clarification of Wormsley's words means that the crossfire stays on Wormsley, on interpreting "THERE ARE BLACK PEOPLE IN THE FUTURE." At the same time, any talk about policing which Picker has set in motion, about racist demands and racist neighbours, is deferred. Although the racism behind the removal was addressed during subsequent discussions, the question of race was always coupled with a reminder of Wormsley's good behaviour. The community offended from Wormsley's message was never put on trial, she was. On white terms and for white peace of mind, the billboard had to be framed as "good behaviour," nothing further.

This is not exactly the case. Wormsley explains that "THERE ARE BLACK PEOPLE IN THE FUTURE" began as her "response to the *absence of non-white faces* in science fiction films and TV." She continues: "[t]he work has become an archive of information, histories and myths that continue *the diaspora's apocalyptic narrative*" (alishabwormsley.com, emphasis mine). The recognition entailed in apocalypse is that of Black people's destruction: Black people are murdered, erased, removed, barred, and driven out of the world. Saying that they have a future is resistance against their ongoing destruction, and a critique of that destruction. Exposing textual and physical erasure, Wormsley challenges white imagination and interrupts dominant discourse. In this sense, "THERE ARE BLACK PEOPLE IN THE FUTURE" acts against those East Liberty residents who find such pronunciations and prospects offensive. It is a 36-foot-long affront on whiteness, not a harmless affirmation a Black woman says to herself in the mirror. On the contrary, the billboard speaks to the past and present which power its inscription. Wormsley explains, and I quote in full:

> I first developed "There Are Black People in the Future" in 2012. I had just moved back to Pittsburgh from New York for residency with the Andy Warhol Museum, and I had a studio. They put me at Westinghouse Academy, which is the middle school and high school in Homewood. I had a classroom there that was like my studio, and kids could come in, and teachers brought

their classes to my class to do workshops. I had approximately 20 regular students that I saw every week making projects. At the time I was making these kind of intense, short experimental sci-fi films, and I was making them in Homewood, actually. And so I got my students to make some, and we would walk around Homewood looking for locations. And the kids would go, "Oh, this is the perfect place for a zombie film, because *it looks like an apocalypse.*" Or, "this is perfect for my *end-of-the-world* film." And I thought, "You know, that's not awesome, because *this is where people live. This is where you live.*" So we started really breaking that down. Like, why does this neighborhood look like this, and why don't other neighborhoods look like this? And what's *the history of this neighborhood?* And how could there be so many amazing people, on the wall of fame, and this neighborhood is here? And what about other Black neighborhoods in America, and what are the similarities? And so we just started really going in, and the kids started making documentary films about it and all kinds of cool stuff. *It's all connected that there are these murders and the prison industrial complex and all these things that I'm talking about with these kids.* ("There Are: An Interview," emphasis mine)

What Wormsley tells the kids is quintessentially different from what Ames tells his son in *Gilead.* Echoing Wormsley in another father-to-child communication, Ta-Ne-hiSi Coates writes:

To be black in the Baltimore of my youth was to be naked before the elements of the world, before all the guns, fists, knives, crack, rape, and disease. The nakedness is not an error, nor pathology. The nakedness is the correct and intended result of policy, the predictable upshot of people forced for centuries to live under fear. The law did not protect us. And now, in your time, the law has become an excuse for stopping and frisking you, which is to say, for furthering the assault on your body. But a society that protects some people through a safety net of schools, government-backed home loans, and ancestral wealth but can only protect you with the club of criminal justice has either failed at enforcing its good intentions or has succeeded at something much darker. (*Between* 17–18)

The "something darker" secured bright futures for some – those who fit in Ames' image – and denied them to others (literally). Like *Stop and Frisk*, the billboard removal aimed to make white residents feel safe and simultaneously assaulted and disarmed Wormsley's body (of work). It should be noted that after Wormsley's message was defused, Picker allowed its reinstatement. Yet, Picker allowed the reinstatement because people like the work, not because others do not (remember, this is the reason why she wanted the billboard taken down). In other words, Picker did not welcome back the billboard because it challenges racist views and bothers racist neighbours; not because it works as disruption to their anti-Black visions and desires (*and it worked!*). Only after the message was rephrased as a harmless affirmation, could Picker concede to what Wormsley says. What becomes clear is that the policing of Black people's demands organizes not only the desire to unsee them, but their conditional acceptance: always already on white terms for "recognition and incorporation are generically anti-Black" (Wilderson, "We're Trying").

6 Beloved endings*

Not all characters are saved for a brighter future. In *Beloved*, about Beloved, Morrison writes:

> They forgot her like a bad dream. After they made up their tales, shaped and decorated them, those that saw her that day on the porch quickly and deliberately forgot her. It took longer for those who had spoken to her, lived with her, fallen in love with her, to forget, until they realized they couldn't remember or repeat a single thing she said, and began to believe that, other than what they themselves were thinking, she hadn't said anything at all. So, in the end, they forgot her too. Remembering seemed unwise ... So they forgot her. (274–275)

Morrison's words hold true not only within the parameters of the story. If we look at Beloved's reception, we will see that a striking number of analyses, those by white literary critics, are premised on leaving Beloved behind (Broeck, "Trauma").[1] Sacrificed for narrative closure and future-orientated readings, Beloved vanishes in a "straightforward tale of feminine community heroics" (242). The persistent focus on reconciliation, or what Broeck calls trauma kitsch in *Beloved*'s reception, eclipses the novel's "excess of non-realism, of non-comprehension, and of non-closure at its narrative (and ethical) core" (248). Instead, white criticism and white writing (to go back to Row) highlights the positive news in the story: the community is no longer disturbed by the trouble at 124 (*it is quiet now, saying nothing*), and Denver might even go to Oberlin – a shining star of radicalism if I can borrow Gilead's description. For many white critics, this is a promising resolution. With Beloved's disappearance, narrative conflict disappears; the closure entails the hope of absolving past traumas:

> the challenge [Morrison] lays down is the work of recovery, of relinking the sundered lines of our mythic past, or reconnecting our present "selves" to the origins from whence we all descended, moving past the restrictions and impositions of ideology toward a common humanity (Tally 140).

---

1 I follow Broeck here and speak to one particular tradition in *Beloved*'s criticism, or what Broeck has identified as trauma kitsch in the reception of *Beloved* (see Broeck's "Trauma"). To such "redemption-focused" criticism can be counted readings like Tally's, as well as John Rohrkemper's "The Site of Memory," Naomi Morgenstern's "Mother's Milk and Sister's Blood," Emilia Ippolito's "History, Oral Memory and Identity", etc. It should be noted that there are other interpretations and analyses of Morrison's text, and that they differ from both "redemption-focused" readings and from Afropessimist ones. Pausing on *Beloved*, I attempt neither thorough nor comparative analyses of its reception (this lies beyond the scope of my study) and this chapter should be read with this in mind.

https://doi.org/10.1515/9783110799996-007

From the excursus on Robinson, we know that such happy ending did not pan out. Despite a Black child like Denver making it to town, she is still denied a future, barred at an unbridgeable threshold. Arriving in *Home*, another Black child pauses before the door. Imagining him standing there, Jack's Black son, Glory realizes the foreclosure of this child's journey. He can be a "figure of inspiration … but not actually a member of her human family," he can arrive but never really enter (Row 112). Incorporating Black characters, Robinson redoubles Glory's curtailed welcome. As if considerate of white readerly communities, Robinson stops Black characters there, as figure-props-raw material, disallowing their presence in the future. We know the latter troubles white vision. The complaints over Wormsley's billboard express white difficulty with Black people "spreading across the landscape" (Row calls this difficulty the "guiding phobia of white American civic life" [124]). Like Picker, Robinson opts for a peaceful resolution; all (images of) Black futurity are removed.

[jumpcut]

From the perspective of Afropessimism, neither Wormsley's, nor Morrison's communications are as positive as white incorporation makes them out to be. Wormsley's refusal to reinstate the billboard is not positive, and Beloved's impossibility to stay is not positive. Not coming back does not mean they have not said anything at all. No hangs in big letters, unremitting: "if she thought anything, it was "No. No. Nono. Nonono. Simple" (Beloved 163). Beloved's passages with disintegrating, abstract language and Beloved's character disappear together with and in the "shaped and decorated" tale. We know the tale has a beginning, middle, and an end and that "the desire for reconciliation is built into the structure of the narrative itself" (Row 7). The tale begins with Sethe and Denver tormented, moves to their conflictual encounter with Paul D, and ends with them being saved. Sethe can put her story next to Paul D's, Denver might go to college, and 124 is quiet (*Beloved* 263–273). This is a tale about forgiveness, recuperation, and hope; the tale is cohesive and linear.

[jumpcut]

From the perspective of Afropessimism, Morrison warns us that the tale camouflages. Beneath it, buried, is "the always now" of white violence because white people simply "don't know when to stop" (B 210, 104). What Morrison calls "the always now" resurfaces in Black critique (*resurfacing, it sounds not like Morrison's individual aesthetic choice, a phrase, a technicality one can remove when troubling*). The always now appears as "the Atlantic now" in Derek Walcott's *Omeros* (129–130), as "the end without ending" in Fred D'Aguiar's *Feeding the Ghosts* (27), as the "afterlife of slavery" in Saidiya Hartman's *Lose Your Mother* (6), as notime "piling up" in Eduard Glissant's *Poetics of Relation* (207). For Wilderson, the condition of negating Black Being (or turning Black Being into nothing) "remains constant, paradigmatically, despite changes in its "performance" over time" (*Red, White, and Black* 75). Orlando Patterson calls the condition a *"permanent violent domination"* (13, emphasis in original); George Wolfe agrees: "nuthin never really ends" in *The Colored Museum.* M. NourbeSe Philip writes: "[i]t is happening always – repeating always, the repetition becoming a haunting" (203). Haunting, Beloved is mistaken for a ghost. Unlike ghosts, however, haunting is "the relentless remembering and reminding that will not be appeased by settler society's assurances of innocence and reconciliation." Eve Tuck and C. Ree continue: the haunting is "a confrontation that settler horror hopes to evade" (642).

[jumpcut]

The reconciliatory tale continues without Black Being. "Every attempt to emplot the slave in a narrative," notes Hartman, "result[s] in his or her obliteration" ("The Position" 184). At the end of *Beloved*, Beloved is forgotten. Yet, to "call the absented thing ghostly suggests a minimal materiality that does not grasp the sheer weight (or lack thereof) of what absence ultimately implies" (McDougall, "Left Out"). Beloved was *there* (Murillo 103–104), so when Wilderson writes that "we need a new language of abstraction to explain" the horrors of anti-Blackness (*Red, White, and Black* 55), it is Beloved who flashes; then **not**. Wilderson explains:

> civil society is held together by a structural prohibition against recognizing and incorporating a being that is dead, despite the fact that this being is sentient and so appears to be very much alive. (41)

Not a ghost but a **negated** Being, Beloved is perhaps the character who tells us this, *look and look again. The cut in the tale, as critique.*

[jumpcut]

Not all critics focused on *Beloved*'s happy, romantic ending.[2] For Dennis Childs, the novel contains passages and a position that are "forward haunting." Childs writes:

---

2 which is, in fact, not *Beloved*'s actual ending. After Paul D returns to Sethe and sees Denver hopeful about the future, Morrison writes another two pages: about Beloved. The last word in the novel is "Beloved"; the two concluding pages are abstract and spoken from outside the story (274–275).

the "forward-haunting" aspects of Morrison's ostensibly past-obsessed text also suggest that gothic penal architectures such as the chain gang cage are not ready to be memorialized in transportation museums as emblems of a bygone era of white supremacy and nascent southern capitalism – that is, if such memorialization within the context of white supremacist culture can amount to anything more than a disqualification of the survivals of unfreedom. Beloved underscores that the terror modalities of chattel slavery have not only survived the putatively static borderline of 1865, but have in fact reached their apogee with the "Security Housing Units" and "Supermaximum" security prisons of today's prison industrial complex. ("You Ain't" 274)

The "survivals of unfreedom" happen now, always; not contained in Sethe's past and unlikely to end with the ultimate gesture of love: cutting Beloved's throat. *To call the cut a gesture of love requires unthinking.* Sethe explains: "if I hadn't killed her she would have died and that is something I could not bear to happen to her." She continues: "[m]y plan was to take us all to the other side where my own ma'am is" (200, 203). Sethe cuts Beloved's throat to save her from death – the death imposed on her one way or the other, in "permanent permutations" (Wolfe *The Colored*). It is this death which will destroy Beloved's body and turns Black bodies into Nothing (Wilderson, *Incognegro* 265). Afropessimistically,[3] this is why Sethe manoeuvres corporeal death (Murillo, 95–120). From here and to go back to Wilderson and Sexton and Sun Ra, the "other side" works as "a sanctuary", it is where Black life can be lived "underground, in outer space", where "[the Black] body changed into something else" and thus force shut the grammar of the world.

[jumpcut]

Within the Western, Cartesian understanding of Self, Sethe's act seems like madness at best, murder at the minimum. Following this understanding, Sethe is sent to prison and rejected from the community. What she does, the cut of the throat, seems too offensive to the world. Morrison writes: "more important than what Sethe had done was what she claimed" (164). Could it be that leaving the world, saying **No** to it, is doing away with its logic and structure, cutting away from them, cutting them to pieces (Broeck, "Thingification")? "To blow the colonial world to smithereens is henceforth a clear image within the grasp and imagination of every colonized subject" writes Fanon (*The Wretched* 6). Moten says:

I believe, [Fanon] comes to believe in the world, which is to say the other world, where we inhabit and maybe even cultivate this absence, this place which shows up here and now,

---

3 to borrow Paula von Gleich's use of the word here (*The Black Border* 11)

in the sovereign's space and time, as absence, darkness, death, things which are not (as John Donne would say). (*The Undercommons* 137).

And then Alexis Pauline Gumbs:

> the first time i thought of you, you were swimming, towards you, through me. first time i thought i was drowning in a world that needed you in it or it would disappear. first time i knew you existed the rest of the history of the world popped like a bubble unready unworthy and my body wanted only future, only you. the first time i felt you move we were deep under-water under something built to keep us under and i couldn't see anything but I understood there was something above everything. above everything despite everything I would find fresh air and breath again. above everything despite everything I would free you. my best idea yet. (13)

Not a "slave" – this is what Sethe decides for herself and her children.

[jumpcut]

Coming back, Beloved disrupts everything. She distorts things we take for normal; makes them seem absurd, even surreal. If Beloved loves Sethe why does she tor-ture her? If Beloved dislikes Paul D, why does she sleep with him? If Beloved is sick and weak, why does she look healthy and grow bigger? Other than absurd and surreal, things look ludicrous for what they really are. Within the logic of whiteness, motherhood and womanhood are denied to Black women, as are kin-ship and sexual pleasure (Spillers, Sharpe, Morrison). No pretence of things getting better can hide the fact that the Negation of Black Being matures: "the terror mo-dalities of chattel slavery have not only survived" but improved. From this perspec-tive, Beloved returns to confront the world, to ridicule and reproach it. Coming back, "Beloved and *Beloved* warp time" (Murillo 102). Coming back, they affront, forcing us to ask – are we not skipping something in the hurry to celebrate recon-ciliation?

> redemption, as *a narrative mode,* [is] a parasite that
> [feeds] upon [Black being] for its coherence
>
> (Wilderson, *Afropessimism* 16).

Who wins from pathological (as Barbara Ehrenreich would call it) optimism, from the desire to always frame the positive side of things? Without optimism, angry, Beloved demands a place in the world. The more she insists that it sickens her, the more this world polices her. Morrison tells us – Beloved is a dead, a dying child. Morrison tells us how the world sees her – gigantic and threatening, too "of-

fensive and divisive" so the world wants to remove her, to silence her objection –
for the sake of white peace that is always already disturbed by Black Being.

[jumpcut]

But, **No.** "No. No. Nono. Nonono." (163) Not a ghost, Beloved haunts.
> so she stood steeple-straight in the shards still in proud
> shock of her quick work. the living room now rainbowed
> glass collected on the floor, the mosaic of how mad she had
> been. she almost laughed out loud but she was too breathless
> at this room, more broken than she. no picture frame could
> contain her now. she was prism reborn. she was sharp refracted
> everything. (Gumbs 29)

Newness

7 Newness and negativity in the northern history of the new

Overview

In the following, I highlight some of the techniques used by white scholars in the conceptualization of renewal. Having discussed rewording, repetition, and over-writing – techniques involving the addition and reiteration of *white content* – I now move to the practice of omission in relaying renewal's story as white history. To do this, I analyse two non-fictional texts – Michael North's 2013 *Novelty: A History of the New* and *The American Jeremiad* by Sacvan Bercovitch, republished in 2012. Both texts thematize renewal and can be understood as stories about it. *Novelty* offers a historical account of theories and examples of newness with the aim to summate them to a somewhat general definition of the new. More critical towards the concept of renewal is *The American Jeremiad* in which Bercovitch explores how the rhetoric of newness animates American culture. For Bercovitch, the idea of the new is part of American mythology which is why *The American Jeremiad* can be read as opposing, at least in intentionality, *Novelty*. Unlike North, Bercovitch does not examine the kinds of newness different discourses and cultural producers have left throughout history but rather investigates how the idea of the new has influenced these discourses and cultural producers, and has moved throughout time.

In this sense, *Novelty* and *The American Jeremiad* represent two quite diverging takes on renewal. In the following, I will show that despite their dissimilarities both texts perform similar agnotological manoeuvres and thus conceal the already discussed duality renewal – negativity. By agnotological manoeuvre, I refer to Robert Proctor's idea of omitted information, or "knowledge that could have been but wasn't, or should be but isn't" (vii). According to Proctor, "ignorance should not be viewed as a simple omission or gap, but rather as an active production" (9). I thus read the omissions in *Novelty* and *The American Jeremiad* symptomatically, and consider to what effect these omissions influence conceptualizations of renewal. I am specifically interested in the ways both texts position renewal in an antagonistic relationship to the past, and conceal the other side of the coin – renewal in relation to destruction, to the production of nothingness.

Informing my analysis with Afropessimist critique, I propose that rather than counter the past, the concept of renewal North and Bercovitch discuss emerges in opposition to negativity. The latter is quintessentially different from the past. For, everything in the past was once new, it once was. Or, the past signifies newness that

https://doi.org/10.1515/9783110799996-008

has previously been; newness that is no longer; the past demarcates existence even though this existence has expired. In contrast, negativity indicates that which is barred from being. Talking about negativity,[1] I follow the Afropessimist contention that 1) Blackness is negated on the level of ontology and reduced to no/thingness (through murder, erasure, theft, derogation), and 2) the invention of this negativity has the function of prolonging white dominance (white renewal) in the process of destroying the Black body as nothing (see Wilderson, Warren). As Wilderson sums up, "there are no Blacks in the world, but, by the same token, there is no world without Blacks" (*Afropessimism* 40).

According to Afropessimism, the modern white subject and the modern world bounce off negativity to come into being. It is the negation of Blackness that counterbalances white property to (re)create, transcend, and fully be in this world. This is the duality I have in mind when I talk about renewal and negativity, and which goes buried in juxtapositions between newness and the past. For, as long as we think of the new in relation to its earlier pronunciations, we find ourselves in a zero-sum game – comparing being to itself, and treating it as sustainable and forever-coming.

To understand renewal in the context of Black negation, i.e. to connect it to white violence, one needs to exfoliate the positive rhetoric through which we define the new, and to reassess its coordinates, its categorization. As it becomes clear from analysing *Novelty* and *The American Jeremiad*, this is a difficult and disagreeable endeavour. Renewal is imbued with hope, progressivism, and morality, and to bring forth its role in the violence that gave birth to the modern world, involves the re-thinking and deconstruction of the positive value of these concepts. As recent critique of Afropessimism reveals (see Okoth, Barlow Jr., Haider, Gordon, et al), any such project is quickly dismissed as fatalistic and unproductive even though Afropessimism exposes the genocidal effects of, for example,

> anti-Black policies on reproductive rights
> the structural elimination of Black life
> deadly living conditions in general
> (F l I n t I s s t I l l w I t h o u t c l e a n w a t e r)

In the following, I examine some of the ways *Novelty* and *The American Jeremiad* uphold a positive understanding of renewal. I look at omissions in the texts as a

---

1 Following Roland Judy, Frank Wilderson III, and Calvin Warren, I use "negativity" to signal both the violence that renders Blackness negative, as no/thing, and the violence that negates Black being (in ontology), see Judy's *DisForming the American Canon* (63–98), Wilderson's *Red, White and Black* (esp. 40–42), and Warren's *Ontological* (esp. 28–50)

manoeuvre that enables ignorance of the relationship between renewal and negativity, and at reading practices that always already instil white content/conceptu-
alization with positivism. I thus join a theoretical turn in knowledge production
that aims to critique and subvert long-standing traditions of knowing and retelling
(hi)story. Here I think as much of interventions like M. NourbeSe Philip's *Zong!* and
Beloved's character, as much of the refusal to obey the constrictions of theoretical
traditions – refusal evident in the work of Fred Moten, Saidiya Hartman, Christina
Sharpe, and Afropessimists like Frank Wilderson III, Calvin Warren, Patrice Douglass, Jared Sexton, and others.

Northern newness, a signpost

In *Novelty*, North makes clear that newness as a human interest and concern is not
only ancient old but overwhelmingly expansive. It is necessary then to curb the
term for the purposes of this analysis. Firstly, North suggests that newness becomes
an absolute value during modernity:

> Whenever it is supposed to start, modernity is always marked off from whatever comes be-
> fore by the conversion of the new from a relative term into an absolute value ... Before nov-
> elty could become "the essence of modernity," it had to be essentialized itself, a host of novel
> effects and stylistic surprises, natural wonders and scientific discoveries, inventions and in-
> novations gathered up into one abstract category explaining and exalting them all. (16)

North continues this temporalization of newness with a remark on Ezra Pound's
paradigmatic dictum to *make it new.* Naturally, references to Pound might direct
us to modernism rather than modernity. After all, Ezra Pound is a key represen-
tative of early modernism and indexing newness's definition with his motto
might situate it there. Yet, North does not say "modernism" and leaves the start
date of modernity explicitly open. It is, of course, impossible to determine whether
North's use of modernity is similar or categorically distinct from mine.[2] Important
to the following analysis is not aligning timeframes but that North treats newness
as a modern (rather than modernist) value at all.

Furthermore, North treats newness as cumulative. Although *Novelty* includes
numerous theories, discourses, historical figures, and works pertinent to newness,
it does not outline a stable, monosemic definition of it. In this regard, North remains faithful to Adorno's contention that "the new is necessarily abstract" (26).
Herewith, I want to pre-empt my analysis with expanding, or reducing, newness

---

2 In my discussion, "modernity" indicates history since the mid 1400's

to such an abstract and absolute category. I will show that in *Novelty* this abstraction and absolutism are realized in their strictly white configurations, and exclude, quite conveniently, key moments in the history North claims to summarize. I begin, therefore, with North's terms, and open the definition of newness as I move along.

A crucial premise for *Novelty* is that newness exists, that it inhabits being. This sounds perhaps as too simple and obvious a proposition but affords the very grounds for North's thematization. Although trajectories of newness seem contested and incomplete, North centres the text where newness appears and uses its known history, proponents, models, sources, and media to construct the category. With this, North joins Niklas Luhmann in suggesting that the very existence of newness is premised on a difference between what something once was and what it will be (159–161). Thus, the new appears as an "ontological nonsense" in that it comes to be "although and because it is *not* what was before" (19, emphasis in original).

How newness presents itself remains uncertain. As North writes, newness' presence might occur as an actual product (60–87), a metaphysical understanding (15–45), or even an uncertainty as it did for postmodernism (144–182). Yet, presence always already marks newness' meaning/essence and adds a specific value to it that bears articulation, questioning, and recognition. In that sense, newness is not only an ontological nonsense but "an ontological possibility" (Novelty 3); it is *something, somewhere*. As Audrey Wasser derives from Derrida, the question "what is?," and *Novelty* intends to answer it for newness,

> is the paradigmatic formula of the metaphysical question of essence … not only does the form of the question attribute a predicable being, essence, or presence … but it also attempts to grasp this being as stable and self-identical. (13)

Later, Wasser follows Maurice Blanchot to conclude that the "essence of a thing can be identified with its condition of possibility" (65). Without going into too much detail,[3] let me signpost *Novelty*'s use of newness in relation to being, presence and possibility: for North, the new is an absolute and abstract value that is indicative of "ontological possibility." Thus, North reaches the conclusion that a "genuine novelty … is major disturbance in the universe, a development like consciousness or life itself." (5) This statement schemes *Novelty* and marks the operative words for North's analysis of newness – disturbance, development, consciousness, and life. I will briefly describe where North finds this kind of newness and then move to the question where he does not as this establishes the more interesting moment in

---

3 I engage this question in my next project, Contours/Conceptualizations of Emptiness.

Novelty (obviously because I think there is a pattern and a purpose to North's omissions, and that they too work for the conceptualization of newness perhaps just as much as North's readable entries).

Progressing presence

To say it in one breath, North tells the Western orthodox story of how newness was understood throughout the ages. He begins with the ancient Greeks, moves through Christianity, touches on Euro-American scientific breakthroughs and technological advances, refers to theories of innovation, and of course, discusses aesthetic modernism and postmodernism. Such linearity on itself contributes to staging newness as accelerating through history, because even though North shows that it, and our understanding of it, have been reliant on models that repeat over periods, its conceptualization is still bound to chronology and Western time.

In fact, cognitive linguists and psychologists draw a connection between writing/reading systems and the ways people imagine temporality. As Fuhrman and Boroditsky show, "the mental time-line extends from left to right for English speakers" ("Mental Time-Lines"), where left stands for past and right for future. Moreover,

> time advances throughout the narrative ... Events that are described later in the text occur after those that are described earlier, unless some linguistic device informs us otherwise. The tense of the narrative is independent of whether the narrative purports to be fact or fiction, and is also independent of the narrative's reference time. (Almeida and Shapiro, "Deictic Centers")

Of course, North's choice to tell the story of newness chronologically might simply be a matter of structuring the text. I want to stress, however, that comprehending the different periods and conceptualizations of newness also happens inside a narrative – *Novelty* itself is a story in which "time advances." If everything "which comes before the now-point [of reading] is in the past," as Almeida and Shapiro write, and "everything that comes after the now-point is in the future from the perspective of that moment of the story," then newness to which *Novelty* is dedicated and which organises it as a meaningful text, reads as advancing. It is not an a- temporal and isolated concept that we understand in each of its fixed to period and discourse meanings because North narrates a story. In it, newness and human interpretations of and endeavours towards it appear as progressing with each page.

The point bears repeating. Let's say, chronological order locates Plato in the first chapter and Pound in the last and North gives an account of how they theorized newness. Because "time advances" in our reading, Plato and Pound's theorizations do not appear as separate constituencies (it is not important whether Pound referenced or considered Plato's work). In fact, reading Plato and Pound in a narrative framework and as part of the same story, means that the first event of newness (Plato) is not split from the last (Pound) but that they are bound through temporality occurring with reading practice.

Western chronopolitics already places Plato's work as prior to Pound's. In reading the narrative, however, time passes for events of newness as well (because they are part of the story). In *Novelty*, a new artwork or a conception of newness does not appear and remain in an isolated period (ancient Greece or modernism) but moves as the story progresses: one idea of newness stays behind (in the past) when another is introduced. Thus, North's newness itself seems to evolve. Put otherwise, we are not reading separate and insulated conceptualizations but a story in which the newness from ancient Greece (mentioned at the beginning) moves through the text, at some point reaching Pound's.

Taking Pound as an example here is cumbersome. As North writes, Pound openly discloses "the full complexity of Make It New ... a dense palimpsest of historical ideas about the new" (165) and never claims the phrase as own. In fact, Pound references the Chinese original of "make it new" whenever he uses the slogan. I agree with North that such referencing shows Pound's unwillingness to sever the expression from its genealogy, and his awareness of the paradoxes inherent to reiterations of calls for originality. The point I want to push here is not that newness has been revised and enriched by previous conceptualizations, whether it is a palimpsest or not, but that it appears as bound to a timeline and advancing throughout the text. In other words, we do not read it as a Greek question or Pound's inscription in a way that these two are unrelated or fixed to page and period. Rather, newness itself piles time. Indeed, we understand it as "a development" as North says wherein different conceptualizations move chronologically, from past (ancient Greece) to future (Western contemporaneity). I leave this argument here with the proviso that linear progressivity is embedded in North's descriptions of newness, and that Novelty charges the term it aims to clarify in ways inseparable from writing/reading practices, knowledge production, and the modes through which one imagines the world.

Unsurprisingly, North's resolve to outline the key models of newness in Western history binds him to certain trajectories. Moreover, North suggests that recurrence and recombination "account for virtually every one of the major ways in which novelty has been conceptualized" and that "their basic shapes were established before Plato" (7). Thus, North has little choice but to connect the dots between ancient Greece and Euro-American contemporaneity and to rely on the big names signalling that route. Novelty lists in encyclopaedic fashion dominant discourses (e.g. Christianity), and history's darlings (Parmenides, Darwin, Wittgenstein, Kant, Bergson, Kuhn, Whitehead – this being a random selection). While a discussion of each of North's entries on newness lays beyond the scope of this project and any attempt to fix their common denominator is precarious at best, I want to emphasize that *Novelty* frames newness in a specific light. Whether it will be the problem of evolutionary biology, Christian rebirth, technological advance, or political and artistic revolutions, newness seems to relate human past to human future on the premises of what once was, came to be, and changed in the history Novelty takes into account. North concludes:

> Any truly generative convention, from DNA to the rules of grammar, must come equipped with some version of this and so on. Even if the extension of it into the future is automatic, though, its expansion is not. In the case of DNA, for instance, biologists have shown that future novelty requires the interpretation of the genetic message within the context of development and environment. Future changes in language obviously require the contributions of future speakers speaking in unprecedented situations. But so long as such extensions are not forestalled, then the possibility of novelty exists- and so it is the indefinite postponement of the end that finally justifies our belief in beginnings. (207)

Novelty's finishing lines do not only frame newness as a mode for premising/promising the future. They also suggest that its nature, however problematic or unresolved, proves durable as a concept and capacity maintained by the temporal tension between end and beginning. This tension seems productive because it opens space to imagine movement toward a better state within Christianity, increase of life in biological terms, advance in technology, or transformations of value, meaning, and their relationality within artistic creation. In all these, newness entails presence even if only as a spectre of the possible and even if its realization fails or challenges "our belief in beginnings" now. "Perhaps there is something in the structure of [newness],"[4] writes North, "that might account for the fact that change and continuity lie so close together" (11). The variety of examples North gives of newness works for the same impression – things change, definitions and understandings change, and newness appears as an ever-deferrable constant.

Indeed, newness's connection to the future (to what has not yet come) allows us to take it onboard over time as a "major disturbance" to be solved or answered but which expands with conceptualizations and history. Even theoretical cancelations like Parmenides's "nothing comes from nothing" (qtd. in North 22), or postmodernist scepticism add to its essence. Thus, *Novelty* poses impossibility as only one side to the problem of newness and develops its image inside-out – with its contradictions, inconsistencies, nonsense, theorizations, dismissals, prod-

---

4 Instead of "newness," North uses "novelty" – he writes earlier that "novelty" is just one of the many names of the new (see North 1). I have used "newness" throughout for consistency

ucts, potentialities and so ons generating what possibly could count as new – in the future.

Thus, just as Darwin makes "change the basic constant of biological existence" (61), so does North link differing stages and reflections on newness to a future that is imaginable as a beginning and a destination. North does so on the premise that, despite or due to alterations over time, *something will possibly exist to reflect the pastness of past experience*; that indeed its time has not yet come but advances. It is because time advances and the way we will talk of and understand the new is still in the future, that what expects us on the horizon is development, consciousness and life – something *in* the universe that will *disturb* its rest by its very *being*. In that sense, North's "unprecedented" situation might follow and repeat newness as we know it and it might diverge from it, it might dis/continue its history and transcend history altogether. For instance, this generativity can be a pathological pathos or paradox as it is for Arendt or Adorno (18). Or, it could be something inconceivably different. As North suggests, we could never know because we are here and now and that what happens next is *narratively ahead.* From this vantage point, newness exists if only as its spectral appearance on the timeline between beginning and end, past and future.

Thus, *Novelty* frames newness as a teleological unknown – it seems that the new will outlast any number of proponents, disbelievers, theories, and products. In the capacity to outlive, however, newness solidifies them as historical and active parts of a dynamic universe, wherein past origin/ality entails ever deferrable futures. It is no wonder then that for North newness is "a social phenomenon" (206). The latter indicates that newness is possible and emergent from correlations between already active/activated participants. Or, that it is a production a/effected in the dynamics between those who are present, voiced, enacted, and eligible – and who are already partaking in, and are visible for, history. North forestalls this conclusion with referring to the American sociologist Randall Collins and locating the understanding of newness as interrelated and derivative, that is, as a social phenomenon, in modernity:

> creativity, both intellectual and artistic, has always been a relative thing, a matter of "new combinations of ideas arising from existing ones, or new ideas structured by opposition to older ones." Novelty ... is also a matter of networks, so that an individual "in isolation rarely develops a new issue or a new way of resolving it." For Collins, this is just as true of ancient Greece as it is of contemporary Manhattan; but this very insight is obviously conditioned by a modern experience, long used to thinking of invention as a diffuse event shared among many equally responsible inventors. (150)

There are two statements I want to unpack here. Firstly, the existence of any social grouping implies that there is at least one outsider who, by not belonging, marks the outlines of the whole. Put bluntly, the concept of being "in" infers that someone is "out" – "in" and "out" are defined in the distinction from each other. As Miranda Joseph writes, social groupings, or communities, "seem inevitably to be constituted in relation to internal and external enemies and that these defining others are then elided, excluded, or actively repressed" (xix). In the example above, a group of "equally responsible inventors" shares sociality and the capacity to (re)create the new. I am yet to determine whom North excluded from this group – who is not "in" on the responsibility and inventiveness. Before I name the excluded, however, I want to repeat that *Novelty* portrays newness as contingent on the production and productivity between already active participants.

One reason for this is embedded in North's reading of history. Despite the meticulous enlisting of examples of newness, *Novelty* is a genealogical account with all the limits genealogical accounts carry. As Russ Castronovo shows,

> genealogy works doubly: it incorporates many daily acts into the historical record, but it ultimately makes this inclusive gesture in order to exclude those persons it deems non-historical. (11)

As it will become evident later, the division between historical and non-historical both precedes and exceeds *Novelty*. For all North's detailed citations, he chooses to exclude Black cultural producers and Black cultural production from the text. This exclusion follows a longstanding tradition of seeing Black bodies as non-historical. As Bethwell Ogot sums up,

> According to … imperial historiography, Africa had no history and therefore the Africans were a people without history. [Imperial, or white, knowledge production] propagated the image of Africa as a "dark continent." Any historical process or movement in the continent was explained as the work of outsiders, whether these be the mythical Hamites or the Caucasoids. (71)

The second thing I want to mention is that, for North, consciousness of this specific characteristic of newness: it being a social phenomenon, develops throughout modernity. Newness is "conditioned by a modern experience." Or, the modern inventors North has in mind, recognize agents who have shared in creation; they have a knowledge of newness's palimpsestic nature and can discern its contributors. This is important because, as mentioned earlier, modernity bears witness to another crucial invention – that of negative and negated Blackness animating, by opposition, white authority (Wilderson, Warren). Or, as Black critique shows, modernity's *New Man* – the active and abled agent of historical progress and its futural perspectives – emerges with modernity's incapacitation of Black bodies, the "always now" of "violence [that] repositions the Black as a void of historical movement" (Wilderson, *Red, White, and Black* 38).

I come back to these points later. For now, I want to signpost that North imbues newness with past, possibility and presence – newness appears as the product of those active and abled inventors who participate in historical progress and recognize, or decide, who is "in" on it. Moreover, by keeping newness abstract and changing – that is, by keeping possibility and presence open to the future – North makes newness's very prospect perform as material enough to reflect and cache meanings and value, even if neither the new as a structure, nor we as future speakers are there to decode them. Or as North says, possibility and presence recur or find different combinations to appear as newness's image. It might be a postponed, obscure, even unattainable one, but its promise flashes in the distance.

Not in the frame

Do you remember where we are?
No way where we are is here

(Moten, "Blackness" 743)

Novelty is, of course, egregiously insular. Of the nearly three hundred names North spells out in full, eleven are female. All included women are white, seven of them American, and all are cited in passing. Only two intellectuals of colour make the cut – Salman Rushdie and P.R. Masani. North names zero Black men. North names zero Black women. It strikes one as odd. Back in 1994 North published on Claude McKay, Jean Toomer, and Zora Neale Hurston and forgetting his deliberations on the Harlem Renaissance cannot be a simple oversight. After all, North concluded that the appropriation of Black work revived the art and careers of white forefathers, Pound among others (*The Dialect* 67–68). He also talked of language transformation in ways that render literary modernism, in its racialized traditions, more than a fitting space to analyse renewal and the new. Moreover, North writes that

> One of the reasons discussions of American language and culture always ended up as discussions about race is that truly original American art forms – jazz, vaudeville, the movies – were created by blacks and dominated by black impersonators ... Logically, then, *the new American writers would be black*, for, as Calverton says, "*In respect of originality...the Negro is more important in the growth of American culture than the white man.*" This is not exactly what the white avant-garde had in mind. Though they were often quite happy to predict great things for black writers in the future, for the present these *[Black] materials and cultural creations would remain raw material for white writers to use.* (*The Dialect* 135, emphasis mine)

If *The Dialect of Modernism* has exhausted the issue of creation/politics for North, however, it conveys another unspoken obvious in *Novelty* – improvisation. The Oxford English Dictionary defines the latter as

> the action or fact of composing or performing music, poetry, drama, etc., spontaneously, or without preparation … The action or fact of doing anything spontaneously, without preparation, or on the spur of the moment; the action of responding to circumstances or making do with what is available.

Improvisation means creating something suddenly; coming up with an action that arises from "what is available." In other words, improvisation figures as the quintessential performance of newness, if not as its immediate source. Perhaps because of this, jazz would have been a justified question in *Novelty*.[5] I say this fully understanding that North could have never anticipated Fred Moten's *In the Break* or Fumi Okiji's *Jazz as Critique.* Yet, academic and popular discourses almost unanimously place jazz at the heart of improvisation. In the words of Burton Perreti, jazz "revitalized improvisation in the music of the Western world" (113).

Like Perreti and despite scientific racism, a staggering number of scholars study jazz as "one of America's original art forms" (Starr and Waterman 20) and as apotheosis of creative power and originality. A "conspicuous feature of modernity" (Rasula 157), jazz surfaces inevitably in questions of modernist art, experimentalism, and the new.[6] North himself links jazz and improvisation in *The Dialect of Modernism* (153), and for many, improvisation constitutes the very "essence of jazz" (Baker 5).

I need to mention here that I do not aim to map out the vast and multidirectional aspects of jazz. Nor can I do justice to jazz' various formal and political developments. I want to stress, however, that analyses of improvisation almost always include, if not accentuate, jazz and jazz- related topics. Even when they do not study jazz techniques, critics and cultural producers almost unanimously connect jazz to questions of newness and find them to be im- and explicit to it (see Baker,

---

5 Of course, jazz is not a singular absence in *Novelty.* Others are perhaps less obvious – like a mention of Jean Michel Basquiat when North references Andy Warhol, of David Blackwell when North discusses information theory, or of the Black Arts Movement when North studies modernism. A comprehensive correction to these omissions pertains to a different project. I choose to stay away from it because I disagree that a simple balancing out of *Novelty's* white and male corpus can drastically change its discursive momentum. After all, *Novelty* is neither the first, nor the last account that forgets Black participation in history, and it is also not original in its views on Western progressivity.

6 Search engines list jazz first in entries for "improvisation," connecting it directly to modernity and the new.

Santi). This is not a marginal perspective. Attention has been paid not only to the emergence of jazz, but also to the political urgencies it reflects, and to the ways these two reverberate in its constant expression of origin/ality and the new. In this sense, conceptualizations of newness underlay both the performance and the analysis of jazz. As Fumi Okiji suggests, jazz figures as and at the primal scene of creative activity:

> A negotiation of the desire to share in the tradition and the imperative to remain distinct is the where the work of a jazz musician is centred. This mimetic attitude is a feature of all artistic pursuit, but it comes into sharp focus when considering jazz. The mimetic negotiations in jazz and other collaborative practices may also be their "unity- constitutive" moments. (9)

Okiji writes further that jazz is not only "capable of reflecting critically on the contradictions from which it arises [but] that it is compelled to do so" (5). Yet, what are the critical reflections inherent to jazz? One crucial reflection is tied directly to newness and creation but not in the sense that newness and creation appear in *Novelty*. To best explain this, let us begin with the connection between Blackness and jazz.

Alain Locke once wrote that "[t]he Negro, strictly speaking, never had a jazz age; he was born that way" (qtd. in Cambridge 169). Like Locke, Okiji and Moten connect jazz to metaphysics and Blackness – and this is the connection I would like to consider here. Let us follow Locke and first examine the second part of his statement, the birth of "the N-." For Hortense Spillers, this birth occurs with the violence altering the Human-ness of Africans transported to Europe during the Middle Passage (*Black* 212). This alteration, Spillers argues, can be best understood as the dehumanization of the African by reducing her to "a thing, to being for the captor" (206). Here is how Franz Fanon describes the negative meanings written into "N-":

> The Negro is an animal, the Negro is bad, the Negro is wicked, the Negro is ugly; look, a Negro; the Negro is trembling, the Negro is trembling because he's cold, the small boy is trembling because he's afraid of the Negro, the Negro is trembling with cold, the cold that chills the bones, the lovely little boy is trembling because he thinks the Negro is trembling with rage, the little white boy runs to his mother's arms: "Maman, the Negro's going to eat me." (*Black Skin* 93)

Drawing on Spillers and building the argument further, Calvin Warren explains that "[s]omething *new* emerges with the transport of the African. The African becomes black-being... utter alterity (metaphysical nothing)" (*Ontological* 38, emphasis mine). Warren concludes:

(1) The Negro is the incarnation of nothing that a metaphysical world tries tirelessly to erad-
icate. Black ~~being~~ is invented precisely for this function ontologically; this is the ontological
labor that the Negro must perform in an anti-Black world. (2) The Negro is invented, or
born into modernity, through an ontometaphysical holocaust that destroys the coordinates
of African existence. The Negro is not a human, since ~~being~~ in not an issue for it, and instead
becomes "available equipment." (27)

For Afropessimists like Warren, the positionality of metaphysical negation, or onticide as Warren calls it, is that of Blackness insofar that Blackness is invented in
and for the transfiguration from African to "N-" (27–62). On the violence of this
invention, Wilderson writes

This violence which turns a body into flesh, ripped apart literally and imaginatively, destroys
the possibility of ontology because it positions the Black in an infinite and indeterminately
horrifying and open vulnerability, an open object made available (which is to say fungible)
for any other subject. As such, "the black has no ontological resistance in the eyes of the
white man" or, more precisely, in the eyes of Humanity (*Red, White, and Black* 38).

If Blackness is born through such violence and Black bodies are forced to reside in
the terror of cancelled being, could this recognition be articulated in jazz, and furthermore, could jazz make such recognition visible? In other words, is this moment, the onticide, implied in Locke's idea of jazz as the birth of "the N-"?

Although Fred Moten's work is far from Afropessimist, it seems to suggest that
much. Moten writes for example that the connection between "jazz, death, race,
and spirit" is metaphysical (*In the Break* 118), and that the question of metaphysics
inevitably connects to rupture and murder (198–201). Okiji articulates a similar
point when she says that – as black expression and as expression of Blackness –
jazz gives "access to a conflicted subject" where conflict indexes the caesura between "human and inhuman, American and black (African)" (4–6). Or, the caesura
between being (born, new) and ~~being~~ (no/thing) (Wilderson, Warren).

Of course, a **[jumpcut]** is needed to concede to the idea that by not including
improvisation – and its most notable expression in American history and culture,
jazz – North precludes the connections one can draw between newness and negativity. After all, jazz is only one of the themes from which the theme of improvisation could have entered *Novelty.* Yet, North makes no indication of how performances of newness might have played out within its histories, nor does he engage
with any Black production that has surfaced in Western versions of the latter.

We can consider this a simple omission, a technicality, even a genealogical in-
evitability. However, when we highlight *Novelty*'s other exclusions, when we see
them together, the omission of jazz and improvisation begin to look more like induced ignorance, "knowledge that could have been but wasn't, or should be but
isn't" (Proctor vii). The **[jumpcut]** from induced ignorance to the associations sup-

pressed knowledge could have triggered if permitted to speak is not that hard to make

so, let me grant space to another omission in Novelty.

If *Novelty* aims to summarize the big names in Western explorations of newness, it certainly overlooks an important one: that of Christopher Columbus. For now, let's only refer to the omission as curious. After all, North ignores a historical figure canonized as "the first American hero, a man set apart from others of his era," the "original pioneer archetype" (Joyce 138). According to Barry Joyce, history textbooks consistently frame Columbus through the ideology and rhetoric of newness. Joyce writes:

> Columbus' role in opening a new world was considered all the more remarkable in light of what he had set out- and failed- to accomplish, which was to sail westward to Asia. "His idea was not realized by himself," reminded John Lord. "But how much grander was the discovery of a new western continent." It was an event that "opened a new era in the history of man- that it was the dawn of *a new civilization*, higher and more perfect than had yet been known," according to J. Olney. God used Columbus to reveal a new world and for a new people-His people- *The People*. All secular past and present paled in the light of this event ... Columbus was God's instrument. Yet he was also an individual who possessed "genius, energy, and enterprise," a new sort of man set apart from those of the Old World. He was *the new Adam*, first of a long line of independent, energized, enterprising men and women destined to inherit this new world. It took nothing less to initiate "the most important event that has ever resulted from individual genius and enterprise." (136–137, emphasis in original).

The newness-rhetoric framing Columbus has long been called into question, as well as has its function in discarding violence against Indigenous populations. Critics in the line of Paula Gunn Allen, Craig Womack, Ward Churchill, and Vine Doloria have identified the narrative manoeuvres employed to rephrase Columbus's violence as greatness, deconstructed the long- lasting marketing of newness, and recognized its tactical usage in the hi/stories of conquest and colonization. One immediate excuse for omitting Columbus then is a departure from a historical record that venerates him and understands 1492 as a marker of discovery and new beginnings. In other words, a refusal to address what has been canonized as America's greatest event of Newness and a Break with the Old World, can be read as North's critique of the tradition and an escape from the conclusions it dictates. In that sense, *Novelty* interrupts the line of praised forefathers, and by excluding the event and its subsequent representations, voids them from significance and effect. Columbus discovered nothing new so he has no place in the history of newness – would read the exemptive for the omission logic.

This reasoning yields some obvious problems. For one, *Novelty* claims to summarize historical events, figures, and theories in their relation to newness. North searches for its definition in the ways newness has been understood, proclaimed and sought in the past. He does not disqualify theories or innovators because their conception of newness has proven untenable or insufficient.[7] Rather, he combines different interpretations of newness (albeit "one-sided"), and assembles a definition with their multiplicity. It seems safe to assume then that just because Columbus did not discover America *because it was there all along*, does not mean that the century-old narrative of discovery and the "New World" offers little to analyses of newness. On the contrary, texts emergent with conquest and colonization – journals, ship logs, scientific entries, travelogues: the repertoire of the "Age of Discovery" – forms perhaps the richest and most significant archive of what the West defined as new.

From this perspective, stories about Columbus as *"the new Adam"* could have strengthened North's definition of newness, and would have also been logical entries in *Novelty*. After all, the historical moment of discovery is a straightforward occasion for contemplating newness. What is more, the encounter with the "New World" is the moment "by which Europe would [later] identify itself" (Judy 73), and therefore is crucial for the history and set of questions North considers. The problem with the discourse around the "New World" and Columbus is that, thanks to Indigenous and Black critique, Columbus has fallen from grace, he has lost signification as "great," "genius," and "enterprising," and come to denote the much harsher reality of Western transgressions towards the Lucayan, Taino, and Arawak people:

> Columbus did not sally forth upon the Atlantic for reasons of "neutral science" or altruism. He went, as his own diaries, reports, and letters make clear, fully expecting to encounter wealth belonging to others. It was his stated purpose to seize this wealth, by whatever means necessary and available, in order to enrich both his sponsors and himself ... he not only symbolizes the process of conquest and genocide which eventually consumed the indigenous peoples of America, but also bears the personal responsibility of having participated in it. (Churchill 6)

Conceptualized together with the brutality of conquest and colonization, the newness of the "New World" obtains quite a different image. What is more, such conceptualization exposes the fact that Western definitions of newness were contingent on dominant representations and language, and not merely constitutive of them. This recognition remains buried in Novelty, and as we have seen previously,

---

7 For instance, North has no problem referencing Nazis like Richard Wagner.

a more promising image of newness emerges – newness connected to future being and possibility, rather than to colonizers' authority to destroy.

Could omitting Columbus and the "New World" in *Novelty* be attributed merely to North's disillusionment with one particular "hero?"[8] Or, do the exclusions of Black original and innovative work and creators join this omission, thus affording a leeway for celebrating historical progress and renewal as induced and performed solely by white authority? Keeping up with contemporary critical theory, North could not have signposted the discourse around Columbus – not without interrogating to some extent the connexion between destruction (negativity) and newness. The point I want to push here is that the discourses hosting formations of white life-affirming capacities and being, and of Black and Indigenous murder and dehumanization overlap. Reasoning the existence of America's new men cannot be divorced from the destruction of native populations – as Black and Indigenous critique show, the latter were both excused with, and assumed to prove, the first.

Unlike North, Black critics continuously connect newness and the production of negativity. Warren writes:

> Once on European soil (and in the hold of the ship), the African ceases to exist and instead becomes "other," an alteration of humanity. Something new emerges with the transport of the African. The African becomes black ~~being~~ and secures the boundaries of the European self – its existential and ontological constitution – by embodying utter alterity (metaphysical nothing). Metaphysics gives birth to black ~~being~~ through various forms of anti-Black violence, and this birth is tantamount to death or worldlessness [This] birth is death – death as nothing, death as the Negro, death as blackness, death as the abyss of metaphysics. (*Ontological* 38–39)

How would *Novelty* have sounded if newness was not understood as that what succeeds death, as what emerges after old things, tradition, and the past disappear, i.e. if *we do not read newness' story as a linear progression unfolding in ever-renewable futures, but as a chokehold, a simultaneous process, the always new and the always now, the new and no/thing?* What damage does such re-thinking inflict on positive categories – progress, change, transcendence, movement forward? How can hope and morality be theorized in the light of their connection to death and destruction? The continuous attack on Afropessimism in academic circles reveals just how unsettling it is to re-think our beliefs, the good, uplifting stories we tell each other. After all, white being and the white world depend on them. Still, what Jack Halberstam saw in Black critique is and here I join his reading:

---

8 As Ward Churchill writes, Columbus is a symbol, he "transcends himself" standing in for an entire discourse (13).

[Black and Indigenous people] want to take apart, dismantle, tear down the structure that, right now, limits our ability to find each other, to see beyond it and to access the places that we know lie outside its walls. We cannot say what new structures will replaces the ones we live with yet, because once we have torn shit down, we will inevitably see more and see differently and feel a new sense of wanting and being and becoming. What we want after "the break" will be different from what we think we want before the break and both are necessarily different from the desire that issues from being in the break. ("The Wild Beyond" 6)

8 Instead of conclusion*

Summary

Previously, I discussed how North highlights newness as a white property and a positive value. I argued that *Novelty*'s omissions afford a background against which newness is separated from anti-Black violence. By dissociating newness from desires for domination and acts of murder and theft, *Novelty* bars the recognition that such acts have often been executed and excused in its name and expression. Only through this textual deferral could newness be reframed as positive and productive. It is this framing that buries the existing relationships between newness and negativity, and silences alternative visions.

From this vantage point, North's omissions generate a useful absence. This is a familiar story. After all, Black absence is an effect of white vision – even the destruction of Black being disappears in the execution of that destruction (whiteness always already alibies itself out as Afropessimists say). The production of Black absence was never a concealed or blurred actuality (*white authority destroys Black bodies before our eyes, it always has*). As Saidiya Hartman contends, the fact that white people unsee the destruction "is due to the sheer denial of black sentience," of framing Black bodies as nothing (*Scenes of Subjection* 51).

We can say then that *Novelty*'s omissions are not more a matter of oversight than they are of logic. They determine which side of the coin becomes visible, shifting readers' focus away from historical tense/tensions – the "always now" of Black negation wherein white renewal develops. Thus, the omissions give space to an imaginable better future, its image relayed from one white generation to the next (recall Campbell and Robinson).The white story repeats itself, its "yet unresolved unfolding" spelt over and over, in a million different ways (Sharpe 14).

This is not the full story – not if we "sit with the nothingness that the world leaves us" (McDougall, "Left Out"), not if we perceive alternative visions *(think of Morrison and Wormsley's articulations). Then, think of how they get removed erased voided.* I have considered this process of removal, of flipping the coin, in the ways white literature ignores or incorporates Black being. Ignorance, pace Proctor, destroys Black being through omission, erasure, and preclusion. This becomes evident in North and DeLillo's refusal to name Black bodies (of work), and the ways this refusal bars alternative vision. Incorporation destroys Black being by accumulation and rewording, i.e. by a change of meaning and its exploitation for the project of whiteness. I examined this in fictional texts like *Kathy Goes to Haiti* and *Gilead* and the ways Acker and Robinson use Blackness on white terms. Both ignorance and incorporation alter perspective. They produce a type

https://doi.org/10.1515/9783110799996-009

of functional absence – a negation, a negativity against which white authority renews itself.

For the purposes of conclusion, a few summary remarks may be ventured here:

- the always new of white being, creation, and possibility ricochets off, i. e. come into existence in opposition to, "the always now" of structural incapacitation where Black being is held;
- white ways of reading, knowing, and relating the world highlight the first part of the equation, the side of the coin white authority wishes and bets on;
- techniques and registers coalesce when white authority retells the same story. The story precludes Black critique and goes on;
- the white story unfolds as the avant-gardism of whiteness; the avant-gardism of whiteness unfolds the white story;
- white stories use negative Blackness, their plot is the negation of Blackness.

Summary remarks carry their own limitations. It remains unclear, for instance, how and when Black being breaks "the grammar of suffering" (to go back to Spillers and Wilderson), and flees. Afropessimists would say that flight is a structural impossibility, and maybe they are right. Yet, the always new and "the always now" are products of authority, and authority can be disobeyed and undone. I say this far away from Eshel and North's optimism, and not looking at the horizon of some distant future. The future is not everybody's if things stay the same, if (hi) story repeats. As Jaye Austin Williams says, if there is hope for Black flight, it is not hope in its traditional conceptualization:

- not hope as a bright and purifying promise
- not hope as fiction (functional for white characters *Kathy Jeffrey Jack* and authority *Acker DeLillo Robinson* even when hope is only fictive)
- not hope in the abstract, as "some casual, idle wish"

Hope is on the ground, busy, physical. It is work, literal (against romanticization and illusion and next-generation-bailouts), and it is work, hard. Hope is dirty, resolute, radical, and stubborn and real. Hope is a verb: *shedisobeys shedestroys* ("Die Welt")

In a certain light, the conflation between hope and dissent suggested above will show its brighter side. I hurry to halt.

I will do so by pausing at another critical study on the concept of renewal – *The American Jeremiad* by Sacvan Bercovitch. Among other things, Bercovitch considers the rhetoric and ritual of newness and the ways they configure the image of America as "the country of tomorrow" and the American "way of life [as] 'futurity' itself" (xx). Tracing its roots to the New England Puritan jeremiad, Bercovitch deconstructs the rags to riches story embedded in American ideology. He writes

> [The] rhetoric of promise as threat, doomsday, and millennium entwined – [the] vision of America as an unfolding prophecy – became ... the foundational national story ... the success story became intrinsic to the narrative of the community at large, and, in time, came to define the nation. Catchwords like "upward mobility" flowered into moral ideals ... only in the United States among modern nations was the self-made man made symbol of a rising nation. That symbol, fusing representative individualism and historical progress, found its territorial correlative in the symbolic newness of the New World ... the New World held the promise of the future. (xiii-xiv)

Bercovitch explains further that the jeremiad, "the colonists' first literary innovation," became a long-lasting cultural phenomenon, one that influenced concepts of Americanness and "opened into an interactive network of art, economy, value system, and public ritual" (xii). The stories of Jeffrey, Kathy, and Jack (and DeLillo, Acker, and Robinson for that matter) could be read against this background. In no small way, all these stories derive from and partake in American mythology. These are the stories of American heroes, who, despite the failings of their Humanly existence, move forward. As the story goes, movement forward unfolds not only as historical progress but as the possibility for a (better) future. In its essence, this is the simple retelling of the American Dream.

Like Bercovitch, I wanted to study how the story is told, to deconstruct it. Mostly because, like him, I am sceptical towards "a redemptive American newness" (xix) and the kind of hope it generates. Unlike Bercovitch, I do not believe in the possibility of finding myself in "a willful no-man's land" (xxiii), nor that the ideology he critiques empowers and constrains everyone in the same way.[1] Assuming the latter, Bercovitch makes several imaginative manoeuvres which speak to the difficulty of critical endeavours, and to the possible traps and failings of scholarly projects.

I will address some of these issues in *The American Jeremiad*, although I refrain from an extensive analysis.[2] Instead, I would like to go back to an idea I began with – the image of work as happening in the factory, a process blocked behind walls that inevitably determine what we produce and how we see the world. Or, our vision(s) are not only limited by the set of rules we follow – ideology as

---

1 Unlike Afropessimists who consider the antagonisms between Black and white positionality as structural, Bercovitch sees them as *"within the culture"* (158).

2 As I mentioned earlier, a comprehensive correction to Bercovitch and North's omissions pertains to a different project. I choose to stay away from it because I disagree that a simple balancing out of omitted information can drastically change these books' discursive momentum.

Bercovitch posits – but by our own positionality *we are never in "no-man's land" even when we work to understand, or dismantle, the ideology which constrains us.*

Take *The American Jeremiad* for instance. Bercovitch deconstructs the American Dream – the success story built into Americanness and the image of America. Yet, the process of such deconstruction can be as dismissive of *destructive to* Black being as the story itself. Like North, Bercovitch omits Black critique. Ta-Nehisi Coates writes

> I wanted to escape into the Dream, to fold my country over my head like a blanket. But this has never been an option because the Dream rests on our [Black] backs, the bedding made from our [Black] bodies. (*Between* 11)

Similarly, Calvin Warren writes that the "American dream ... is realized through black suffering." Warren continues: "it is the humiliated, incarcerated, mutilated, and terrorized black body that serves as the vestibule for the Democracy that is to come." ("Black Nihilism" 6). Back in 1936, Langston Hughes wrote: "(America never was America to me)" ("Let"). Almost thirty years later, Malcolm X declared: "I don't see any American dream; I see an American nightmare" ("Ballot or Bullet Speech"). And yet, for Bercovitch, the American Dream works for everyone, equally:

> Blacks and Indians too could learn to be True Americans, when in the fullness of time they would adopt the tenets of black and red capitalism. John Brown could join Adams, Franklin, and Jefferson in the pantheon of Revolutionary heroes when it was understood that he wanted to fulfill (rather than undermine) the American dream. On that provision, Jews and even Catholics could eventually become sons and daughters of the American Revolution. On those grounds, even such unlikely candidates for perfection as Alaska, Hawaii, and Puerto Rico could become America. (160)

Bercovitch subsumes Black struggles for survival, Black critique, into an Americanness that supposedly promises futurity to everyone. None of the stories I discussed earlier rehearse such concept of Americanness. On the contrary, these stories simultaneously exclude and capitalize on Blackness, render it a useful material for cohesion, and remove its image from the future "that is to come" (*think of Wormsley's billboard*).

Incorporating Black critique, Bercovitch erases it. Thus, he concludes that "the grievances of repressed groups ... reaffirm the American Dream, their faith in the national promise [carry] with it a wonderful vitality" (xxv). Can we misread Coates and Warren's accounts (and Afropessimism) as *faith* in the system? If Calvin and Warren are too recent an example, what about the radical 1960s and 1970s which, according to Bercovitch, inspired *The American Jeremiad?* Was there no Black critique then that questions the system and the official record of (hi)story?

Or, was there not, along that long unfolding of American progress, Black and Indigenous opposition that urged to destroy the system, to break the grammar of the world?

[jumpcut]

In 1952, we read Ralph Ellison:

> I am an invisible man. No, I am not a spook like those who haunted Edgar Allen Poe; nor I am one of your Hollywood-movie ectoplasms. I am a man of substance, of flesh and bone, fiber and liquids – and I might even be said to possess a mind. I am invisible understand, simply because people refuse to see me. Like the bodyless heads you see sometimes in circus sideshows, it is as though I have been surrounded by mirrors of hard, distorting glass. When they approach me they see only my surroundings, themselves, or figments of their imagination – indeed, everything and anything except me. Nor is my invisibility exactly a matter of a biochemical accident to my epidermis. That invisibility to which I refer occurs because of a particular disposition of the eyes of those with whom I come in contact. A matter of the construction of the *inner* eyes, those eyes with which they look through their physical eyes upon reality. (3)

[jumpcut]

In 1978, Audre Lorde:

> The difference between poetry and rhetoric
> is being ready to kill
> yourself
> instead of your children.
> I am trapped on a desert of raw gunshot wounds
> and a dead child dragging his shattered black
> face off the edge of my sleep
> blood from his punctured cheeks and shoulders
> is the only liquid for miles
> and my stomach
> churns at the imagined taste while
> my mouth splits into dry lips
> without loyalty or reason
> thirsting for the wetness of his blood
> as it sinks into the whiteness
> of the desert where I am lost
> without imagery or magic
> trying to make power out of hatred and destruction
> trying to heal my dying son with kisses
> only the sun will bleach his bones quicker.
> A policeman who shot down a ten year old in Queens

stood over the boy with his cop shoes in childish blood
and a voice said "Die you little motherfucker" and
there are tapes to prove it. At his trial
this policeman said in his own defense
"I didn't notice the size nor nothing else
only the color". And
there are tapes to prove that, too.
Today that 37 year old white man
with 13 years of police forcing
was set free
by eleven white men who said they were satisfied
justice had been done
and one Black Woman who said
"They convinced me" meaning
they had dragged her 4'10" black Woman's frame
over the hot coals
of four centuries of white male approval
until she let go
the first real power she ever had
and lined her own womb with cement
to make a graveyard for our children.
I have not been able to touch the destruction
within me.
But unless I learn to use
the difference between poetry and rhetoric
my power too will run corrupt as poisonous mold
or lie limp and useless as an unconnected wire
and one day I will take my teenaged plug
and connect it to the nearest socket
raping an 85 year old white woman
who is somebody's mother
and as I beat her senseless and set a torch to her bed
a greek chorus will be singing in 3/4 time
"Poor thing. She never hurt a soul. What beasts they are."

("Power")

[jumpcut]

And now, John Murillo:

[BLACK PEOPLE] WORK with the shards of Black life and death that called out to [them] because [they] knew and know that the critical, caring, and perilous work [they] need to do is
bound up with destruction. These fragments of Black life and death surrounding [them] af-
firmed [their] sense of [their] own untimeliness against the neatness of time, and of [their]
own stankiness in the middle of nowhere. The untimeliness that signals [Black] destructive
relationship to human models and experiences of time and the stankiness that signals

[Black] destructive relationship to human spaces and spatiality act as the Black *prima materia*, the Black and essential material, with which we must work to create these impossible stories [Black people] imagine, witness, bear, conjure, and live in and against the anti-Black cosmos where and when [they] cannot be. What [Black people] knew, and now know with excruciating intimacy, to be the violent, distorted fabric of spacetime shaping the field of fragments around [Black people] is the material [they] must bend to create Black pocket universes from streets to pages (and everywhere and when between). [Black people] knew and know that in order to conjure Black spacetimes that might upend the anti-Black cosmos, [they] would have to become avatars of destruction, able to bend the forces of untimeliness and stankiness and love toward the kinds of authentic upheaval that must be born if [they] are to save the earth and conjure the impossible story of a wholly unimaginable world. (185)

[jumpcut]

These are rhetorical questions. In this study, I listed a few of the Black thinkers who critique and resist white American (hi)story, the way it tells itself, and befalls.

[jumpcut] *I think of white (hi)story as fallen upon us, like night. It consumes everything around us and conceals the counter-histories, the "insurgent, disruptive narratives that are marginalized and derailed before they ever gain a footing." Yet, someone calls and someone sees, and someone is out there doing the work.*

This critique continues to be ignored, subsumed, and reworded in white interpretations. "Oppressive language," Morrison writes, "does more than represent violence; it is violence" (*What Moves* 201). For Bercovitch, Black critique sounds just like any other articulation of dissent, it appears to use the same language, the same repertoire. He says

> Shortcomings from various "inalienable rights" ("granted by the Founding Fathers") [are] denounced through variant reiteration of those rights, such as the African American Declaration of Sentiments, the Seneca Falls Declaration of Women's Suffrage, the Gettysburg Address, and during the sixties and seventies ceremonial readings of the Declaration itself. (xviii)

What Bercovitch forgets is that Black thinkers have had to find ways out of what was given and there (the rights and expressions Bercovitch considers were a priori anti-Black and so Black work cut away from them). Misseeing the cut, Bercovitch subsumes white and Black perspectives and positionalities and thus makes Black being disappear. Ignoring the incommensurability between Black and white work, Bercovitch can then claim that diversity criticism is just another part of the American ritual (xxiii). He can say that "the United States [has] developed as a country with less diversity in terms of competing ideological alternatives" (xxii).

Yet, the alternatives have always been there.
Removed Alternatives is Not the Same Thing as Alternatives Not Being There.
Someone is out there, doing the work.[3]

---

3 I want to end on this note – rather than closing this book, the note allows me to open another one.

9 Acknowledgements

I will end this book as I should have started it. Black critics continue to theorize and untie the connections between newness and negativity, and do so with unprecedented originality, against the lures and limitations of this world – just read M. NourbeSe Philip's *Zong!* or *Propter Nos* or King et al.'s *Otherwise Worlds* and see for yourself. [1] As I have written elsewhere, it will be not up to me, a white reader, to interpret these texts.[2] Instead, I have tried to listen, to see how Black work dismantles the knowledge I have – given and taken in a specific light, setting. I do not know whether an escape is ever possible, but I believe that one can question, search for a way out, crumble or climb inside themselves.

I hope that beyond an academic study, this book lives up to what it is – something personal, a meditation. Sure, I wanted to see how white texts frame renewal as a strictly white capacity and prerogative – this story runs through literature just as it unfolds daily, dictated by dominant discourse. I also wanted to take *Minor Cosmopolitanisms*[3] literally – to where minor may be singular but significant, and cosmopolitanisms an interplay of ideas. Thus, I could ask – how do we read in the privacy of our bias, knowledge, and traditions? What movements do we make – involuntary, conscious, visible like choices and invincible like sight in full darkness? Do we pause, re-consider direction? Books give, my mother says, not only the stories they contain but the stories they do not. This is what I wanted to share – the process of looking into, the story I managed to put down and the space it leaves for another one.

I am aware that many of the white writers I discussed here were on the lookout for the same things – whether on theory or in practice – and so I also know that the structure of the quest is flawed, and our capacities – limited. For that, and many other reasons, this book was difficult to write. My supervisors know. I am thankful for the rock-hard patience and generous support and what I can only call love with which they calmed my thunders and stood by me. It would have been impossible to finish the book without them – so, thank you, wholeheartedly.

An earlier version of the chapter on DeLillo appeared in the 2019 Special Issue of *COPAS*, "White Supremacy in the United States and Beyond." I also presented drafts of the chapter at a Symposium in Potsdam in 2019 and at *House of World*

---

1 *Propter Nos:* esp. Volume 3
2 See *Of the Fugue in the Passage to Madness*
3 The research for this monograph was generously funded through a PhD fellowship at the Research Training Group *Minor Cosmopolitanisms* (October 2016 – March 2020)

https://doi.org/10.1515/9783110799996-010

Cultures (Berlin) in 2018. I presented the draft chapter on Acker at the *Research Seminar Series* (UNSW, Australia) in 2019. An early version of the Acker chapter was also published in *The Minor on the Move. Doing Cosmopolitanisms.* The poems springing from, related to, and reflecting on this project were published in *MUZ2: Insel, Tote und Touristen* in 2022. Further reflections on the project were published in *Minor Cosmopolitan: Thinking Art, Politics and the Universe Together Otherwise* and *U.S Studies Online: Forum for New Writing.* I thank all editors, readers, reviewers, co-panelists, and conference participants for their feedback and support.

The research for this monograph was funded through a PhD fellowship at the Research Training Group *Minor Cosmopolitanisms* (October 2016 – March 2020). The RTG funded also my participation in conferences, workshops, and a winter/ summer schools in and outside Europe, as well as the extensive research stays in Australia in my pursuit of a joint PhD degree. I am thankful to everyone I met during this journey and for the support and guidance I received. Prof. Franco Barchiesi, Dr. Anke Bartels, Prof. Anne Brewster, Prof. Sabine Broeck. Dr. Michal Buech, Prof. Bernd Boesel, Dr. Judith Coffey, Prof. Sergio Costa, Prof. Lars Eckstein, Dr. Cedric Essi, Dr. Paula von Gleich, Dr. Marius Henderson, Prof. Rozena Maart, Dr. Anouk Madoerin, Dr. Taija McDougall, Dr. Courtney Moffett-Bateau, Dr. Moses Maerz, Dr Laetitia Nanquette, Prof. Sean Pryor, Prof. Alan Rice, Prof. Anja Schwarz, Dr. Sikho Siyotula, Dr. Samira Spatzek, Prof. Jens Temmen, Prof. Nicole Waller, Prof. Dirk Wiemann, Prof. Jaye Austin Williams, Prof. Zairong Xiang, and the activist and artist Roxley Foley and Gustavo Mendez Lopez: I listened to the best of my ability, with gratitude, always. I also thank Prof. Carsten Junker, Prof. Julia Roth and Prof. Darieck Scott for accepting my manuscript for their *American Frictions* series, and the in-house editor Dr. Julie Miess. Several women at the social network *WIASN* and at the law firm *Goldenstein Rechtsanwälte* offered friendship and encouragement – my thanks go to them, too. My family is a gift which no earthly gratefulness can acknowledge – you have all my love. Of course, all mistakes and insufficiencies in the present study remain mine. The friends and family who re-lived and soothed every sign of exhaustion and desperation, who talked when I needed someone to talk, and listened when I needed someone to listen, who cooked and cheered, and beyond anything held me, strong – this book would have been impossible without you. I hope you all know how much you mean to me, not only on this path, not only on this page.

Hugs!

Marrri

Works cited

Aaron, Daniel. "How to Read Don DeLillo." *Introducing Don DeLillo.* Duke UP, 1991.

Acker, Kathy. "A Conversation with Kathy Acker by Ellen G. Friendman." Dalkey Archive Press. https://www.dalkeyarchive.com/a-conversation-with-kathy-acker-by-ellen-g- friedman/. Accessed April 2020.

Acker,Kathy."Against Ordinary Language: The Language of the Body." https://www.yvonnebuchheim.com/uploads/1/7/0/8/17088324/acker- kathy_the_language_of_the_body.pdf. Accessed April 2020.

Acker, Kathy. *Hello, I am Erica Jong.* Contact II Publications, 1988.

Acker, Kathy. *Kathy Goes to Haiti.*1978. Flamingo, 1993.

Acker, Kathy. *Bodies of Work: Essays.* Consortium Book Sales & Dist,1997.

Acker, Kathy. *Blood and Guts in High School.* Grove Press, 1984.

Acker, Kathy. "Few Notes on Two of My Books." *Review of Contemporary Fiction*, 31 – 36. Dalkey Archive Press, 1989.

Acker, Kathy. *Great Expectations.* Grove Press, 1983.

Acker, Kathy. "Interviewed by Deaton." *Textual Practice.* Vol. 6, Issue 2, 1992. https://doi.org/10.1080/09502369208582142. Accessed August 2019.

Acker,Kathy."Interview by Mark Magill." Bomb Magazine # 6, 1983. http://bombmagazine.org/article/307/kathy-acker. Accessed April 2020.

Acker,Kathy."Interview with McRobbie at the ICA." *Youtube.* https://www.youtube.com/watch?v=Pxu85kiqkg0. Accessed March 2020.

Acker, Kathy. *My Death My Life by Pier Paolo Pasolini.* 1984. Grove Press, 1988.

Acker, Kathy. "Requiem." 1998. *CTHEORY: Act III* https://journals.uvic.ca/index.php/ctheory/article/view/14632/5498. Accessed April 2020.

Acker, Kathy. "The Gift of Disease." 1997. *Guardian Newspapers Limited* https://kathyackerlibrary.tumblr.com/post/65518036995/the-gift-of-disease. Accessed April 2020.

Adams, James. *The Epic of America.* 1931.Transaction Publishers, 2012.

Adas, Michael. *Machines as the Measure of Men.* Cornell UP, 2014.

Adorno, Theodor. *Aesthetic Theory.* 1970. Bloomsbury, 2013.

Afrofuturist Abolitionists of the Americas. Wordpress. https://afanarchists.wordpress.com/2019/10/12/anarkata-a-statement/. Accessed May 2020.

Aldridge, Taylor. "Black Bodies, White Cubes: The Problem with Contemporary's Art Appropriation of Race."

ArtNews. https://www.artnews.com/art-news/news/black- bodies-white-cubes-the-problem-with-contemporary-arts-appropriation-of-race-6648/ Accessed January 2020.

Allen, Paula Gunn, et. al. *Columbus and Beyond: Views from Native Americans.* Western National Parks Association, 1992.

Allen, Theodor. *The Invention of the White Race: The origin of racial oppression in Anglo-America. Volume II: The Origin of Racial Oppression in Anglo-America.* Verso, 1997.

Almeida, Michael, et. al. "Deictic Centers and the Cognitive Structure of Narrative Comprehension." Researchgate, June 2013. https://www.researchgate.net/publication/237251965_Deictic_Centers_And_The_Cog nitive_Structure_of_Narrative_Comprehension. Accessed April 2020.

Arendt, Hannah. *The Origins of Totalitarianism.* 1966. A Harvest Book, 1976.

https://doi.org/10.1515/9783110799996-011

Armstrong, M.C. "Back in the USSR: Returning to History in Don DeLillo's Zero K." *The Mantle*, June 2018, www.themantle.com/literature/back-ussr-returning-history-don-delilloszero-k. Accessed 20 June 2019.

Bailey, Lisa M. Siefker. "Fraught with Fire: Race and Theology in Marilynne Robinson's Gilead." *Christianity and Literature* 59.2. 265 – 80. Sage, 2010.

Baker, David. *Jazz Improvisation (Revised): A Comprehensive Method for All.* Frangipani Press, 2005.

Barlow Jr., Michael A. "Addressing Shortcomings in Afropessimism." *InquiriesJournal* 8.09 (2016). http://www.inquiriesjournal.com/a?id=1435>. Accessed April 2020.

Baucom, Ian. *Specters of the Atlantic, Finance Capital, Slavery, and the Philosophy of History.* Duke University Press, 2005.

Baxter, Charles. "A Different Kind of Delirium." *The New York Review of Books*, 9 Feb. 2012, www.ny books.com/articles/2012/02/09/different-kind-delirium/. Accessed 26 Sept. 2019.

Bell, Derrick. *Faces at the Bottom of the Well: The Permanence of Racism.* Basic Books, 1992.

Benjamin, Walter. *The Work of Art in the Age of Mechanical Reproduction.* 1935. Penguin, 2008.

Benston, Kimberley. *Performing Blackness: Enactments of African-American Modernism.* Routledge, 2013.

Bercovitch, Sacvan. *The American Jeremiad.* 1978. The University of Wisconsin Press, 2012.

Berlant, Lauren and Kathleen Steward. *The Hundreds.* Duke University Press, 2019.

Bhabha, Homi K. "The White Stuff." *Artforum International.* Vol. 36 No. 9. May 1998.

Black Agenda Report. News, Commentary and Analysis from the Black Left. https://www.black agendareport.com/. Accessed April 2020.

Black, Hannah. "'The Painting Must Go': Hannah Black Pens Open Letter to the Whitney About ControversialBiennial Work."*Artnews*, March2017. https://www.artnews.com/art-news/news/the-painting-must-go-hannah-black- pens-open-letter-to-the-whitney-about-controversial-biennial-work-7992/.Accessed April 2020.

Blain, Keisha. "There are Black People in the Future": An Interview with Artist Alisha

Blanchard, Pascal, et al. *Human Zoos: Science and Spectacle in the Age of Colonial Empires.* Liverpool UP, 2008.

Bogle, Donald. *Toms, Coons, Mulattoes, Mammies, and Bucks: An Interpretive History of Blacks in American Films.* Continuum, 2001.

Brady, Nicholas. "Louder Than the Dark: Towards an Acoustics of Suffering." *Academia.Edu.* https://www.academia.edu/2776515/Louder_than_the_Dark_Towards_an_Acoustics_of_Suffering. Accessed March 2020.

Brand, Dionne. *A Map to the Door of No Return: Notes to Belonging.* Random House, 2001.

Brand, Dionne. *In Another Place, Not Here.* Knopf Canada, 2011.

Bodomo, Nautama and Frances Bodomo. *Afronauts.* 2014. Short Film. https://www.youtube.com/watch?v=Ib3pu5jXWHU. Accessed April 2020.

Borowska, Emilia. *The Politics of Kathy Acker: Revolution and the Avant-Garde.* Edinburgh University Press, 2019.

Broeck, Sabine. "Thingification and Un-homing of Black Refugee Life in European Discourses", forthcoming.

Broeck, Sabine. "Trauma, Agency, Kitsch and the Excesses of the Real: Beloved within the Field of Critical Response." *Plotting against Modernity: Critical Interventions in Race and Gender.* ed. Esders, Haertel and Junker, Ulrike Helmer Verlag, 2014.

Buck-Morss, Susan. *Dreamworld and Catastrophe.* MIT Press, 2000.

Buchloch, B.H.D. *Neo-Avantgarde and Culture Industry. Essays on European and American Art from 1955 to 1975.* The MIT Press, 2000.

Buerger, Peter. *Theory of the Avant-Garde.* 1974. Manchester UP, 1984.

Butler, Octavia. *Dawn.* 1987. Warner Books, 1997.

Cambridge Dictionary. n.d. https://dictionary.cambridge.org/ Accessed November 2022.

Campbell, John. *Negro-Mania: Being and Examination of the Falsely Assumed Equality Of the Various Races of Men.* Campbell & Power, 1851.

Campt, Tina. "Black Feminist Futures and the Practice of Fugitivity." *Youtube.* Oct. 2014. https://www.youtube.com/watch?v=2ozhqw840PU. Accessed April 2019.

Campt, Tina. *Listening to Images.* Duke UP, 2017.

Castronovo, Russ. *Fathering the Nation: American Genealogies of Slavery and Freedom.* University of California Press, 1995.

Césaire, Aimé. *Discourse on Colonialism.* Monthly Review, 2000.

Churchill, Ward. *From a Native Son: Selected Essays in Indigenism, 1985–1995.* South End Press, 1996.

Childs, Dennis. "You Ain't Seen Nothin' Yet": "Beloved," the American Chain Gang, and the Middle Passage Remix." *American Quarterly,* Vol. 61, No. 2, 271–297. John Hopkins UP, 2009.

Clare,Ralph."WhyKathyAckerNow?." *Los Angeles Review of Books.*2018. https://lareviewofbooks.org/article/why-kathy-acker-now/ Accessed April 2020.

Clawson, Rosalee. "Poor People, Black Faces: The Portrayal of Poverty in Economics Textbooks." *Journal of Black Studies,* Vol. 32, No. 3. 352–361, 2002.

Coates, Ta-Nehisi. *Between the World and Me.* Spiegel & Grau, 2015.

Coates,Ta-Nehisi."The First White President." *The Atlantic,* October2017. https://www.theatlantic.com/magazine/archive/2017/10/the-first-white-president-ta-nehisi-coates/537909/. Accessed April 2020.

Colby, Georgina. *Kathy Acker: Writing the Impossible.* Edinburgh University Press, 2016.

Crane, Diana. *The Transformation of the Avant-Garde: The New York Art World, 1940–1985: New York Art World, 1940–85.* University of Chicago Press, 1987.

Crum, Maddie. "The Bottom Line: 'Zero K' By Don DeLillo." *Huffpost.* April, 2016. https://www.huffpost.com/entry/don-delillo-zero-k_n_57291880e4b016f37893feae. Accessed April 2020.

Csicsery-Ronay,Jr., Istvan. "An Elaborate Suggestion: Review of Brian McHale's Constructing Postmodernism." *SFS* 20:457–64, #61, 1993.

D'Aguiar, Fred, editor. *Feeding the Ghosts.* Chatto and Windus, 1997.

Darwin, Francis. *The Life and Letters of Charles Darwin.* John Murray. 1887.

DeLillo, Don. *Zero K.* Scribner, 2016.

Diamond, Elin, editor. *Performance and Cultural Politics.* Routledge, 1996.

Dini, Rachele. "Don DeLillo, Zero K." *European Journal of American Studies,* 14 May 2016, https://journals.openedition.org/ejas/11393. Accessed 1 December 2019.

De Zwaan, Victoria. 1997. "Rethinking the Slipstream: Kathy Acker Reads Neuromancer." *Science Fiction Studies,* Volume 24, Part 3, https://www.depauw.edu/sfs/backissues/73/dezwaan73.htm. Accessed April 2020.

Doloria, Vine. *Red Earth, White Lies: Native Americans and the Myth of Scientific Fact.* Fulcrum Publishing, 2018.

Dos Passos, John. *The Fourteenth Chronicle. Letters and Diaries of John Dos Passos.* Gambit, 1973.

Douglas, Christopher. "Christian Multiculturalism and Unlearned History in Marilynne Robinson's Gilead." *Novel: A Forum on Fiction* 44.3. 333–53, 2011.

Duvall, John. "Introduction: The Power of History and the Persistence of Mystery." *The Cambridge Companion to Don DeLillo.* Cambridge UP, 2008.

Du Bois, W.E.B. B*lack Reconstruction in America.1860–1880.* 1935.The Free Press, 1998.

Du Bois, W.E.B. *The Comet.* Grand Central Publishing, 2001.

Dyer, Richard. *White: Essays on Race and Culture.* Routledge, 1997.

Ehrenreich, Barbara. *Smile or Die. How Positive Thinking Fooled America and the World.* Granta Books, 2010.

Ellison, Ralph. *Invisible Man.*1952.Vintage, 1990.

Ellwood, Johnson. *The Goodly Word: The Puritan Influence in American Literature.* Clements Publishing, 1995.

Engebretson, Alex. *Understanding Marilynne Robinson.* University of South Carolina Press, 2017.

Eshel, Amir. *Futurity: Contemporary Literature and the Quest for the Past.* The University of Chicago Press, 2013.

Fanon, Frantz. *Black Skin, White Masks.* 1952. Transl. Charles Markmann. Pluto Press, 1967.

Fanon, Frantz. *The Wretched of the Earth.* 1961.Transl. Richard Philcox. Groove Press, 2004.

Felder, Cain. editor *Stony the Road we Trod: African American Biblical Interpretation. Fortress Press.*1991.

Felder, Cain. *Original African Heritage Study Bible-KJV.* Judson Press, 2007.

Ferris, Joshua. "Joshua Ferris Reviews Don DeLillo's 'Zero K.'" *The New York Times,* 2 May 2016,www. nytimes.com/2016/05/08/books/review/don-delillos-zero-k.html. Accessed 12 June 2019.

Fiedler, Leslie. "The New Mutants." *Partisan Review.* 1965. http://www.bu.edu/partisanreview/books/ PR1965V32N4/HTML/files/assets/basic- html/index.html#505. Accessed April 2020.

Fields, Karen and Barbara Fields. *Racecraft. The Soul of Inequality in American Life.* Verso, 2014.

Foucault, Michel. *The Order of Things.* 1966. Routledge, 2002.

Friedman, Ellen. "Kathy Acker: Wandering Jew." *Kathy Acker and Transnationalsim,* ed. Kathryn Nicol and Polina Mackay, xi-xxi. Cambridge Scholars, 2009.

Fuhrman, Orly and Lera Borodski. "Mental Time-Lines Follow Writing Direction: Comparing English and

HebrewSpeakers." https://pdfs.semanticscholar.org/70a5/a1fc1bbe823a002d003aeb5d8d503f0be5a7. pdf. Accessed April 2020.

Fusco, Coco. "Censorship, Not the Painting, Must Go: On Dana Schutz's Image of Emmett Till." *Hyperallergic,*March2017. https://hyperallergic.com/368290/censorship-not-the-painting-must-go-on-dana-schutzs-image-of-emmett-till/. Accessed March 2020.

Gilroy, Paul. *The Black Atlantic: Modernity and Double Consciousness.* Verso, 2002.

Greenberg, Clement. "Avant-Garde and Kitsch." 1939. http://sites.uci.edu/form/files/2015/01/Green berg-Clement-Avant-Garde-and-Kitsch- copy.pdf. Accessed April 2020.

Guinn, Clay. "You Get Off on Stealing: Kathy Acker, Travelogues, and the American Imperial Instinct." *Plaza: Dialogues in Language and Literature 3.1:* 13 – 23. 2013 https://journals.tdl.org/plaza/index. php/plaza/article/ Accessed April 2020.

Glissant, Edouard. *Poetique de la Relation: Poetique III.* Gallimard, 1990.

Glueck, Robert. "The Greatness of Kathy Acker." *Lust for Life, on the Writings of Kathy Acker,* ed. Scholder, Harryman, And Ronell. Verso, 2006.

Gordon, Alex. "Updated: "There Are Black People In The Future" text removed from East Liberty public-art project at behest of landlord." *Blogh,* April 2018. https://www.pghcitypaper.com/ Blogh/archives/2018/04/05/there-are-black- people-in-the-future-text-removed-from-east-liberty-public-art-project-at- behest-of-landlord. Accessed 2020.

Gordon, Lewis R. "Shifting the Geography of Reason in Black and Africana Studies." *The Black Scholar,* vol. 50, no. 3, 2020, 42 – 47

Gumbs, Alexis Pauline. *Spill. Scenes of Black Fugitivity.* Duke UP, 2016.

Haider, Asad. *Mistaken Identity: Race and Class in the Age of Trump.* Verso, 2018.

Halberstam, Judith. *The Queer Art of Failure.* Duke Up, 2011.

Halberstam, Jack. "The Wild Beyond." *The Undercommons. Fugitive Planning and Black Study.* Stefano Harney and Fred Moten. Minor Compositions, 2013.

Hall, Stuart. "Race, Articulation and Societies Structured in Dominance." *Sociological Theories: Race and Colonialism.* UNESCO, 1980.

Harney, Elizabeth and Ruth B. Phillips, editors. *Mapping Modernisms: Art, Indigeneity, Colonialism.* Duke UP, 2018.

Hartman, Saidiya. "Memoirs of Return." *Rites of Return.* Ed. Hirsch and Miller. 107 – 23. Columbia UP, 2011.

Hartman, Saidiya. *Scenes of Subjection, Terror, Slavery, and Self-Making in Nineteenth- Century America.* Oxford University Press, 1997.

Hartman, Saidiya. "The Position of the Unthought": An Interview with Saidiya V. Hartman." by Frank Wilderson. *Qui Parle.* Vol. 13. No 2. 183 – 201, 2003.

Hartman, Saidiya. "Venus in Two Acts." *Small Axe.* No 26. 1 – 14, 2008.

Hassan, Ihab. The Dismemberment of Orpheus. Toward A Postmodern Literature.1971. The University ofWisconsin Press, 1982.

Henderson, Margaret. "From Counterculture to Punk culture: The Emergence of Kathy Acker's Punk Poetics." *LIT: Literature Interpretation Theory.* Vol. 26, Issue 4, 2015. https://doi.org/10.1080/10436928.2015.1092347. Accessed March 2018.

Herren, Graley. *The Self-Reflexive Art of Don DeLillo.* Bloomsbury Academic, 2019.

Hinojosa, Lynne. "John Ames as Historiographer: Pacifism, Racial Reconciliation, and Agape in Marylinne Robinson's Gilead." *Religion & Literature,* Vol. 47, No. 2, 117 – 142. The University of Notre Dame, 2015.

hooks, bell. *Talking Back. Thinking Feminist, Thinking Black.* South End Press, 1989.

Hutcheon, Linda. *A Poetics of Postmodernism: History, Theory, Fiction.* Routledge, 1988.

Hong, Cathy. "Delusions of Whiteness in the Avant-Garde." *Lana Turner. # 7,* November 2014. https://arcade.stanford.edu/content/delusions-whiteness-avant-garde. Accessed November 2022

Horkheimer, Max and Theodor Adorno. "The Culture Industry: Enlightenment as Mass-Deception." *Dialectic of Enlightenment.* Continuum, 1982.

Ippolito, Emilia. "History, Oral Memory and Identity in Toni Morrison's Beloved." *The Poetics of Memory.* Tuebingen: Stauffenburg, 1998, 199 – 203.

Jack, Alison. "Barth's Reading of the Parable of the Prodigal Son in Marilynne Robinson's Gilead: Exploring Christlikeness and Homecoming in the Novel" *Literature and Theology,* Volume 32,Issue1,March2018, 100 – 116, https://doi.org/10.1093/litthe/frx018. Accessed April 2020.

James, Joy. "Reaching Beyond "Black Faces in High Places": An Interview With Joy James". *Truthout,* February 2001.

Jillson, Calvin. *Pursuing the American Dream: Opportunity and Exclusion Over Four Centuries.* University Press of Kansas, 2004.

Joseph, Miranda. *Against the Romance of Community.* University of Minnesota Press, 2002.

Joyce, Barry. *The First U.S. History Textbooks: Constructing and Disseminating the American Tale in the Nineteenth Century.* Lexington Books, 2015.

Judy, Ronald. *(Dis)Forming the American Canon: African-Arabic Slave Narratives and the Vernacular.* University of Minnesota Press, 1993.

Kakutani, Michiko. "In Don DeLillo's 'Zero K,' Daring to Outwit Death." *The New York Times,* 25 Apr. 2016, www.nytimes.com/2016/04/26/books/review-in-don-delillos- zero-kdaring-to-outwit-death.html. Accessed 12 June 2019.

Kalmar, Ivan. *White but not quite: Race and illiberalism in Central Europe.* Bristol University Press, 2022.

"Kathy Acker Goes to Haiti: Review." *Goodreads.* https://www.goodreads.com/book/show/146908. Kathy_Goes_To_Haiti. Accessed April 2020.

Kealy, Sean and David Shenk. *The Early Church and Africa.* Oxford University Press, 1975.

Keeling, Kara. *Queer Times, Black Futures.* New York UP, 2019.

Key, Laura and Brittany Noble. *Analysis of Ferdinand's de Saussure's Course in General Linguistics.* Routledge, 2017.

Khalil, Andrea. *The Arab Avant-garde: Experiments in North African Art and Literature.* Praeger, 2003.

Kidd, James. "Don DeLillo's Latest 'Zero K' Is a Visionary Novel about Cryogenics." *INEWS,* 20 May 2016, https://inews.co.uk/culture/books/don-delillos-latest-zero-k-visionarynovel-cryonegics-538202. Accessed 26 Sept. 2019.

King, Tiffany Lethabo, Jenell Navarro and Andrea Smith, editors. *Otherwise Worlds. Against Settler Colonialism and Anti-Blackness.* Duke UP, 2020.

Kocela, Christopher. "A Myth beyond the Phallus: Female Fetishism in Kathy Acker's Late Novels." *Genders.* no. 34, 2001. https://www.colorado.edu/gendersarchive1998-2013/2001/09/01/myth-beyond-phallus-female-fetishism-kathy-ackers-late-novels. Accssed April 2020.

Kraus, Chris. *After Kathy Acker: A Biography.* Penguin, 2018.

Krauss, Rosalind. *The Originality of the Avant-garde and Other Modernist Myths.* 1985. MIT Press, 1986.

Ladner, Gerhard. *The Idea of Reform. Its Impact on Christian Thought and Action in the Age of the Fathers.*1959. Wipf and Stock Publishers, 2004.

Lakoff, George and Mark Johnson. *Metaphors We Live By.* University of Chicago Press, 2008.

Lear, Jonathan. *Wisdom Won From Illness.* Harvard UP, 2017.

Le Gall, Yann. *Remembering the Dismembered: African Human Remains and Memory Cultures after Repatriation.* forthcoming.

Lee, Robert. *Modern American Counter Writing. Beats, Outriders, Ethnics.* Routledge, 2010.

Leise, Christopher. "That Little Incandescence": Reading the Fragmentary and John Calvin in Marilynne Robinson's Gilead." *Studies in the Novel* 41.3 (Fall, 2009): 348 – 67.

Lentricchia, Frank. *Introducing Don DeLillo.* Duke UP, 1991.

Levin, Ann. "Don DeLillo's New Novel Considers Life After Death." *Daily Herald,* 2 May 2016, https://www.dailyherald.com/article/20160502/entlife/305029848. Accessed 1 Dec. 2019.

Lopez, Ian. *White by Law. The Legal Constructions of Race.* New Your UP, 2006.

Lorde, Audre. *The Collected Poems of Audre Lorde.* W. W. Norton & Company, 1997.

Lowe, Lisa. *The Intimacies of Four Continents.* Duke UP, 2015.

Mann, Paul. *The Theory-Death of the Avant-Garde.* Indiana University Press, 1991.

Manners, Marylin. "The Dissolute Feminisms of Kathy Acker." *Future Crossings: Literature Between Philosophy and Cultural Studies*, edited by Krzysztof Ziarek and Seamus Deane, 98 – 119. Northwestern University Press, 2000.

Martinot, Steve, and Jared Sexton. "The Avant-Garde of White Supremacy." *Social Identities,* 9: 2, 169 – 181, 2003. DOI: 10.1080/1350463032000101542 Accessed April 2020.

Mbiti, John. *Bible and Theology in African Christianity.* Oxford University Press, 1986.

Mcbride, Jason. "TheLast Days of Kathy Acker." *Hazzlit.* July, 2015. https://hazlitt.net/feature/last-days-kathyacker. Accessed March 2020.

McDougall, Taija. "Left Out: Notes on Absence, Nothingness and the Black Prisoner Theorist." *Anthurium. A Caribbean Studies Journal.* https://anthurium.miami.edu/articles/10.33596/anth.391/?fbclid=IwAR05Nia- 4GrNbou_fE9_3TvJs8bsdr5QVyib9CkoDWiqF3p1d9Zy3ZN3n7 A. Accessed May 2020.

Mckissic, William. *Beyond Roots: In Search of Blacks in the Bible.* Mckissic Enterprises LLC, 2017.

McKittrick, Katherine. *Demonic Grounds. Black Women and the Cartographies of Struggle.* University of Minnesota Press, 2006.

McLaren, Peter. "Unthinking Whiteness: Rearticulating Diasporic Practice." *Revolutionary Pedagogies. Cultural Politics, Instituting Education, and the Discourse of Theory.* Ed. Peter Trifonas. Routledge, 2000.

Mills, Katie. *The Road Story and the Rebel. Moving through Film, Fiction, and Television.* Southern Illinois University Press, 2006.

Mills, Sara. *Discourses of Difference: an Analysis of Women's Travel Writing and Colonialism.* Routledge,1991.

Merriam-Webster Dictionary. Merriam-Webster. n.d. https://www.merriam-webster.com/. Accessed April 2020.

Mirzoeff, Nicholas. *The Right to Look. A Counterhistory of Visuality.* Duke UP, 2011.

"MLA Formatting and Style Guide." *The Purdue OWL, Purdue U Writing Lab.* Accessed April 2020.

MoMa Online. n.d. https://www.moma.org/collection/terms/165. Accessed April 2020.

Morgenstern, Naomi. "Mother's Milk and Sister's Blood". *Differences.* 8. No 2, 1997, 101–126.

Morrison, Toni. *Beloved.* 1987. Vintage, 1997.

Morrison, Toni. "Interview with Charlie Rose." *PBS,* 1993.

Morrison, Toni. *Playing in the Dark, Whiteness and the Literary Imagination.* Harvard UP, 1992.

Morrison, Toni. *What Moves at the Margin: Selected Nonfiction.* University Press of Mississippi, 2008.

Moten, Fred. "Blackness and Nothingness." *The South Atlantic Quarterly.* 112:4 Fall 2013, 737–780.

Moten, Fred. "Do Black Lives Matter?: Robin D.G. Kelley and Fred Moten in Conversation."*Vimeo.* 6th Jan. 2015. https://vimeo.com/116111740. Accessed April 2019.

Moten, Fred. *In the Break: The Aesthetics of the Black Radical Tradition.* University of Minnesota Press, 2003.

Moten, Fred and Stefano Harney. T*he Undercommons. Fugitive Planning and Black Study.* Minor Compositions, 2013.

Muhammad, Ismail. "On Seeing Blackness. How do you look at images of anti-black violence without reproducing that violence in perpetuity?" *Real Life,* May 2017. https://reallifemag.com/on-seeing-blackness/. Accessed 2020.

Muhlestein, Daniel. "Vision as Creation and Alternative: The Role of the Author Function in Marilynne Robinson's Plural Text Gospels of Gilead." *Irish Journal of American Studies.* http://ijas.iaas.ie/issue-6-daniel-muhlestein/. Accessed March 2020.

Murillo, John. *Impossible Stories: On the Space and Time of Black Destructive Creation.* The Ohio State University Press, 2021.

Muñoz-Alonso, Lorena. "Dana Schutz's Painting of Emmett Till at Whitney Biennial Sparks Protest." *Artnet,*
March, 2017. https://news.artnet.com/art- world/dana-schutz-painting-emmett-till-whitney-biennial-protest-897929. Accessed April 2020.

Muñoz, José. *Cruising Utopia: The Then and There of Queer Futurity.* New York University Press, 2009.

Myles, Lessie. *Discovering Black People in the Bible.* Independently Published, 2018.

Nielsen, Donald. *Horrible Workers. Max Stirner, Arthur Rimbaud, Robert Johnson, and the Charles Manson Circle. Studies in Moral Experience and Cultural Expression.* Lexington Books, 2005.

Nikolova, Mariya. "Of the Fugue in the Passage to Madness. Toni Morrison's Beloved and NourbeSe Philip'sZong! AConversation in Three Parts. " https://www.academia.edu/28926931/Of_the_Fugue_in_the_Passage_to_Madness._T

oni_Morrisons_Beloved_and_NourbeSe_Philips_Zong_A_Conversation_in_Three_P arts. Accessed April 2020.

Nikolova, Mariya. "White Violence and Spectral Blackness in Don DeLillo's Zero K." *COPAS.* Vol. 20, No 2. (2019). http://dx.doi.org/10.5283/copas.320. Accessed April 2020.

Nikolova, Mariya. "Nekrologika." *MUZ 2: Insel, Tote, und Touristen*, p. 96 – 101. Mittel und Zweck, 2022.

Nikolova, Mariya. "Reaching the Limit. Or, How Kathy Acker Used Blackness to Abandon Haiti and Arrive Home Safely." *The Minor on the Move. Doing Cosmopolitanisms*, Edition Assemblage, 2021.

Nikolova, Mariya. "Jumpcut"& "Pinecones." *Minor Cosmopolitan: Thinking Art, Politics and the Universe Together Otherwise*, Diaphanes, 2020.

Nikolova, Mariya. "On Breaking Dissertations, or How I Read Sideways." *U.S. Studies Online: Forum for New Writing*, 22 January 2018, http://www.baas.ac.uk/usso/onbreaking-dissertations-or-how-i-read-sideways/

North, Michael. *Novelty. A History of the New.* The University of Chicago Press, 2013.

O'Driscoll, Bill. "'There Are Black People In The Future' Resident Artists Present Their Projects." *Wesa Fm*, October 2019. https://www.wesa.fm/post/there-are-black-people-future-resident-artists-present-their-projects#stream/0. Accessed April 2020.

OED Online. Oxford University Press. n.d. https://oed.com/ Accessed April. 2020.

Ogot, Betwell. "African Historiography: From colonial historiography to UNESCO's general historyof Africa."file:///C:/Users/marri/Downloads/16429- %23 %23default.genres.article%23 % 23 – 17697 – 1 – 10 – 20150309 %20(3).pdf. Accessed April 2020.

Okiji, Fumi. *Jazz as Critique: Adorno and Black Expression Revisited.* Stanford University Press, 2018.

Okoth, Kevin. "The Flatness of Blackness: Afropessimism and the Erasure of Anti-Colonial Thought." *Libcom*,January2020. https://libcom.org/library/flatness-blackness-Afropessimism- erasure-anti-colonial thought. Accessed April 2020.

Olson, Lanse. "Kathy Acker: Queen of the Pirates." 1997. http://www.lanceolsen.com/queen.html. Accessed April 2020.

Painter, Rebecca M. "Loyalty Meets Prodigality: The Reality of Grace in Marilynne Robinson's Fiction." *Christianity and Literature* 59.2. 321 – 40. Sage, 2010.

Pak, Yumi. "Jack Boughton has a wife and a child": Generative Blackness in Marilynne Robinson's Gilead and Home." *This Life, This World: New Essays on Marilynne Robinson's Housekeeping, Gilead, and Home.* Ed. Jason W. Stevens. Brill, 2015.

Parker, Alan. "Mortal Panic: On Don DeLillo's Zero K." *Kenyon Review*, 2016, www.kenyonreview.org/re views/zero-k-by-don-delillo-738439/.Accessed 1 Dec. 2019.

Parks, Suzan-Lori. "New Black Math." *Theatre Journal*, Vol. 57, No 4. John Hopkins University Press, 2005.

Patterson, Orlando. *Slavery and Social Death: A Comparative Study.* Harvard UP, 1982.

Petit, Susan. "Finding Flannery O'Connor's "Good Man" in Marilynne Robinson's Gilead and Home." *Christianity and Literature* 59.2. 301 – 18. Sage, 2010.

Perreti, Burton. *The Creation of Jazz: Music, Race, and Culture in Urban America.* University of Illinois Press, 1994.

Phelan, Peggy. *Unmarked: The Politics of Performance.* Psychology Press, 1993.

Philip, M. NourbeSe. *Zong!* Wesleyan University Press, 2008.

Pitchford, Nicola. *Tactical Readings. Feminist Postmodernism in the Novels of Kathy Acker and Angela Carter.* Bucknell UP, 2002.

Poggioli, Renato. *The Theory of the Avant-Garde.* 1962. Harvard University Press, 1968.

Pratt, Mary Louise. *Imperial Eyes: Travel Writing and Transculturation.* Routledge, 1992.

Proctor, Robert and Londa Schiebinger. *Agnotology. The Making and the Unmaking of Ignorance.* Stanford University Press, 2008.

Pyke, Susan Mary. *Animal Visions. Posthumanist Dream Writing.* Springer, 2019.

Rabinovitz, Lauren. *Points of Resistance: Women, Power & Politics in the New York Avant-garde Cinema, 1943–71.* University of Illinois Press, 2003.

Rasack, Sherene. *Race, Space, and the Law: Unmapping a White Settler Society.* Between the Lines, 2002.

Rasula, Jed. "Jazz and American Modernism." *The Cambridge Companion to American Modernism.* ed. Kalaidjian. Cambridge UP, 2005.

Ra, Sun. "Space is the Place." *Space is the Place: The Lives and Times of Sun Ra.* John Szwed. Pantheon Books, 1997.

Ra, Sun. "Cosmic Equation." *Sun Ra. The Immeasurable Equation.* ed. Wolf and Geerken. Waithawhile, 2005.

Richards, Jill. *The Fury Archives: Female Citizenship, Human Rights, and the International Avant-Gardes.* Columbia University Press, 2020.

Riley, Denise. *Impersonal Passion. Language as Affect.* Duke UP, 2005.

Riley, Shannon. "Kathy Goes to Haiti: Sex, Race, and Occupation in Kathy Acker's Voodoo Travel Narrative." *Kathy Acker and Transnationalsim*, ed. Kathryn Nicol and Polina Mackay, 29–51. Cambridge Scholars, 1991.

Roberts, John. *Revolutionary Time and the Avant-Garde.* Verso, 2015.

Robinson, Marilynne. *Gilead.* Picador, 2004.

Robinson, Marilynne. *Home.* Virago, 2008.

Robinson, Marilynne. "The Faith Behind the Fiction." *Reform.* https://www.reform magazine.co.uk/2016/05/marilynne-robinson-interview/. Accessed April 2020.

Rohrkemper, John. "The Site of Memory: Narrative and Meaning in Toni Morrison's Beloved" *Midwestern Miscellany 24*, 1996, 51–62.

Rosenberg. Harold. *The Tradition of the New.* Hachette Books, 1994.

Ross, Jack. "Review: Kathy Goes to Haiti." *Travel Writing Anthology.* http://139326anthology.blogspot.com/2016/07/review-kathy-goes-to-haiti-1978.html. Accessed May 2019.

Row, Jess. *White Flights: Race, Fiction, and the American Imagination.* Graywolf 2019.

Said, Edward. *Orientalism.* Vintage, 1979.

Saint-Simon, Claude-Henry. *Opinions littéraires, philosophiques et industrielles.* 1825. Ligaran, 2016.

Santi, Marina, editor. *Improvisation: Between Technique and Spontaneity.* Cambridge Scholars Publishing, 2010.

Saramo, Samira. "The Meta-Violence of Trumpism." *European Journal of American Studies*, vol. 12, no. 2, 2017, doi:10.4000/ejas.12129. Accessed 8 Mar. 2019.

Scafidi, Susan. *Who Owns Culture?: Appropriation and Authenticity in American Law.* 1968. Rutgers University Press, 2005.

Schutz, Dana. *Artnet.* http://www.artnet.com/artists/dana-schutz/. Accessed April 2020.

Sciolino, Martina. "Kathy Acker and the Postmodern Subject of Feminism." *College English 52*, no. 4: 437–45. 1990. doi:10.2307/377661. Accessed January 2020.

Sell, Mike. *The Avant-Garde: Race, Religion, War.* Seagull Books, 2011.

Sexton, Jared. "The Social Life of Social Death: On Afropessimism and Black Optimism" *InTensions Journal.* Issue 5. 2011. https://www.yorku.ca/intent/issue5/articles/pdfs/jaredsextonarticle.pdf. Accessed April 2020.

Sharpe, Christina. "Blackness, Sexuality, and Entertainment." *American Literary History,* Volume 24, Issue 4, Winter 2012, Pages 827–841, https://doi.org/10.1093/alh/ajs046. Accessed March 2020.

Sharpe, Christina. *In the Wake: On Blackness and Being.* Duke UP, 2016.

Sharpe, Christina. *Monstrous Intimacies: Making Post-Slavery Subjects (Perverse Modernities).* Duke UP, 2010.

Sharpe, Christina. "Response to 'Ante-Anti-Blackness.'" *Lateral: Journal of the Cultural Studies Association,* vol. 1, 2012. Doi:10.25158/L1.1.17. Accessed 12 Sept. 2019.

Sharp, Sarah. "Artist's Billboard Declaring "There Are Black People in the Future" Taken Down by Landlord." *Hyperallergic,* April 2018. https://hyperallergic.com/436763/alisha-wormsley-the-last-billboard-pittsburgh- there-are-black-people-in-the-future/. Accessed April 2020.

Shugart, Helene. "Postmodern irony as subversive rhetorical strategy." *Western Journal of Communication,* 63:4, 433–455, 1999, DOI: 10.1080/10570319909374653

Shy, Todd. "Religion and Marilynne Robinson." *Salmagundi* 155/156, 251–64, 2007.

Siyotula, Sikho. *Visualising Southern African Late Iron Age Settlements in the Digital Age.* forthcoming.

Smith, Adrianna. "The Nature of the Horizon." *IJAS Online,* No. 6, Special Issue, pp. 22–31. Irish Association for American Studies, 2017.

Sontag, Susan. *Styles of Radical Will.* Penguin, 2013.

Sostek, Anya. "Landlord: 'There Are Black People in the Future' billboard violated lease agreement." *Pittsburgh Post-Gazette,* April 2018. https://www.post- gazette.com/local/city/2018/04/05/Billboard-There-Are-Black-People-in-the- Future-East-Liberty-Eve-Picker-Alisha-Wormsley-Jon-Rubin/stories/201804050187. Accessed April 2020.

Spillers, Hortense. "Interstices: A Small Drama of Words." B*lack, White,and In Color: Essays on American Literature and Culture.*152–176. The University of Chicago Press, 2003.

Spillers, Hortense. *Black, White and in Color: Essays on American Literature and Culture.* University of Chicago Press, 2003.

Spillers, Hortense. "Mama's Baby, Papa's Maybe: an American Grammar Book." *Diacritics* Vol. 17. No 2, 64–81, 1987.

Stam, Robert and Ella Shohat. *Unthinking Eurocentrism. Multiculturalism and the Media.* Routledge, 1994.

Stansell, Amanda. *"Collage Politics": Experimental Narrative and Collective Identity in the 1930s.* University of Wisconsin-Madison, 2003.

Starr, Larry and Christopher Waterman. *American Popular Music from Minstrelsy to MP3.* Oxford UP, 2010.

Stein, Gertrude. "The Gradual Making of The Making of Americans." *Selected Writings,* ed. Carl van Vechten. Random House, 1946.

Stiles, Kristine. "Never Enough is Something Else: Feminist Performance Art, Probity, and the Avant-Garde." *Contours of the Theatrical Avant-Garde: Performance and Textuality.* ed. James M. Harding, Wisconsin Press, 2000.

Suleiman, Susan Rubin. *Subversive Intent: Gender, Politics, and the Avant-garde.* Harvard UP, 1990.

Sweet, Paige. "Where's the Booty?: The Stakes of Textual and Economic Piracy as Seen Throughthe Work of Kathy Acker." *Darkmatter.* 2009 http://www.darkmatter101.org/site/2009/12/20/where%E2%80%99s-the-booty-the- stakes-of-textual-and-economic-piracy-as-seen-through-the-work-of-kathy-acker/ Accessed April 2020.

Sykes, Rachel. "Those Same Trees: Narrative Sequence and Simultaneity in Marilynne Robinson's Gilead Novels." *Irish Journal for American Studies,* Issue 6. http://ijas.iaas.ie/issue-6-rachel-sykes/. Accessed March 2020.

Tally, Justine. *Toni Morrison's "Beloved": Origins.* Routledge, 2008.

Tanner, Laura E. "Looking Back from the Grave": Sensory Perception and the Anticipation of Absence in Marilynne Robinson's Gilead." *Contemporary Literature* 48.2, 227–52, 2007.

Taoua, Phillis. *Forms of protest: anti-colonialism and avant-gardes in Africa, the Caribbean, and France.* Heinemann, 2002.

Terrefe, Selamawit. "What Exceeds the Hold?: An Interview with Christina Sharpe." *Rhizomes. Cultural Studies in Emerging Knowledge.* Issue 29 (2016), https://doi.org/10.20415/rhiz/029.e06. Accessed April 2020.

The Bible. Authorized King James Version, Oxford UP, 1998.

The Cambridge Companion to Modernism. ed. Kalaidjian. Cambridge UP, 2005.

Thiel, Klaus. "Samisdat – The Secret Message of Sun Ra's Poetry." *Sun Ra. The Immeasurable Equation.* ed. Wolf and Geerken. Waithawhile, 2005.

Thomas, Greg. "Afro-Blue Notes: The Death of Afro-pessimism (2.0)?." *Theory & Event*, vol. 21, no. 1, 2018, pp. 282–317

Towers, Robert. "From the Grassy Knoll." *The New York Review of Books*, 18 Aug. 1988, www.nybooks. com/articles/1988/08/18/from-the-grassy-knoll/. Accessed 12 Sept. 2019.

Tuck, Eve and C. Ree. "A Glossary of Haunting." *Handbook of Autoethnography*, ed. Stacey Holman Jones,

Tony E. Adams, and Carolyn Ellis, 639–658. Left Coast Press, 2013.

Vargas, Joao Costa, and Joy James. "Refusing Blackness-as-Victimization: Trayvon Martin and the Black Cyborgs." *Pursuing Trayvon Martin: Historical Contexts and Contemporary Meditations of Racial Dynamics.* Ed. Yancy and Jones. 193–204.Lexington Books, 2012.

von Gleich, Paula. *The Black Border and Fugitive Narration in Black American Literature.* De Gruyter, 2022.

Walcott, Derek. *Omeros.* Farrar, Strauss, Giroux, 1990.

Warren, Calvin L. "Black Nihilism and the Politics of Hope." *Ill Will Editions.* https://illwilleditions.no blogs.org/files/2015/09/Warren-Black-Nihilism-the-Politics- of-Hope-READ.pdfhttps://illwilleditions. noblogs.org/files/2015/09/Warren-Black- Nihilism-the-Politics-of-Hope-READ.pdf. Accessed April 2020.

Warren, Calvin L. *Ontological Terror: Blackness, Nihilism, and Emancipation.* Duke UP, 2018.

Wasser, Audrey. *The Work of Difference: Modernism, Romanticism, and the Production of Literary Form.* Fordham University Press, 2016.

Weele, Michael Vander "Marilynne Robinson's Gilead and the Difficult Gift of Human Exchange." *Christianity and Literature* 59.2, 217–39. Sage, 2010.

Weheliye, Alexander. *Habeas Viscus: Racializing Assemblages, Biopolitics, and Black Feminist Theories of the Human.* Duke UP, 2014.

Wilderson, Frank. *Afropessimism.* Liveright Publishing, 2020.

Wilderson, Frank. "The Prison Slave as Hegemony's (Silent) Scandal." *Social Justice*, vol. 30, no. 2 (92), 2003, pp. 18–27.

Wilderson, Frank and Patrice Douglass. "The Violence of Presence: Metaphysics in a Blackened World." *The Black Scholar*, Vol. 43, No. 4, The Role of Black Philosophy (Winter 2013), pp. 117–123. Francis and Taylor, 2013.

Wilderson, Frank. *Red, White & Black. Cinema and the Structure of U.S. Antagonisms.* Duke UP, 2010.

Wilderson, Frank. *Incognegro: A Memoir of Exile and Apartheid.* South End Press, 2008.

Wilderson, Frank. "We Are Trying to Destroy the World." I*ll Will Editions.* Interview Transcription, 2014. https://illwilleditions.noblogs.org/files/2015/09/Wilderson-We- Are-Trying-to-Destroy-the-World-READ.pdf. Accessed April 2020.

Williams, Jaye Austin. "Die Welt als Desaster ist Realität." Kritish-Lesen. https://kritisch-lesen.de/inter view/die-welt-als-desaster-ist-realitat. Accessed May 2020.

Winkiel, Laura. *Modernism, Race and Manifestos.* Cambridge UP, 2011.

Winton, Laura. *A Rightful Inheritance: Locating the Black Avant-garde.* Academia.edu, 2009 https://www. academia.edu/28060411/A_Rightful_Inheritance_Locating_the_Black_A vant-Garde. Accessed March 2020.

Wolfe, George. "The Colored Museum." 1986. Video Clip. *YouTube.* 18. Aug. 2009. Web. April 2020.

Wollen, Peter. "Death and (Life) of the Author." *London Review of Books.* Vol. 20, No 3, 1998. https:// www.lrb.co.uk/the-paper/v20/n03/peter-wollen/death-and-life-of-the- author. Accessed March 2020.

Womack, Craig. *Red on Red: Native American Literary Separatism.* University of Minnesota Press, 1999.

Wormsley, Alisha B. https://alishabwormsley.com/ Accessed March 2020.

Wormsley, Alisha B."PublicBooks, November 2019. https://www.publicbooks.org/there-are-black-peo ple-in-thefuture-an-interview- with-artist-alisha-b-wormsley/. Accessed April 2020.

Wynter, Sylvia. "No Humans Involved: A Letter to my Colleagues." *Voices of the African Diaspora* 8:2 (1992). http://readingfanon.blogspot.com/2016/05/sylvia-wynter-no-humans-involved-open.html. Accessed April 2020.

Wynter, Sylvia. "Rethinking 'Aesthetics': Notes Towards a Deciphering Practice." *Ex-Iles: Essays on Caribbean Cinema.* Ed. M. Cham. Trenton. Africa World Press, 1992.

Wynter, Sylvia. "Towards the Sociogenic Principle: Fanon, Identity, the Puzzle of Conscious Experience, and

What It Is Like to Be Black." *National Identities and Sociopolitical Changes in Latin America.* Ed. Durán-Cogan and Gòmez-Moriana. Routledge, 2001.

Wynter, Sylvia. "1492: A New World View." *Race, Discourse, and the Origin of the Americas: A New World View.* Ed. Sylvia Wynter, Vera Lawrence Hyatt, and Rex Nettleford. Smithsonian Institution Press, 1995.

Yancy, George. "Whiteness and the Return of the Black Body." *The Journal of Speculative Philosophy. NewSeries,* Vol. 19, No. 4 (2005), 215 – 241. https://www.jstor.org/stable/25670583?seq=1. Accessed March 2020.

Ziyad, Hari. "Why Do White Liberal Artists Love Black Death So Much?" *Afropunk,* March2017.https:// afropunk.com/2017/03/why-do-white-liberal-artists-love- black-death-so-much/. Accessed April 2020.

Index

https://doi.org/10.1515/9783110799996-012